CHEMOMETRICS AND DATA ANALYSIS IN CHROMATOGRAPHY

Edited by **Vu Dang Hoang**

Chemometrics and Data Analysis in Chromatography
http://dx.doi.org/10.5772/intechopen.72089
Edited by Vu Dang Hoang

Contributors

Vu Dang Hoang, Mercedes G. López, Juan Vázquez-Martínez, A. H. Al-Bagawi, M. S. El-Shahawi, G.I. Mohammed, Z.M. Saigl, W. Ahmad, H. Alwael, Carlos Sanz, Angjelina Belaj, Mar Pascual, Ana G. Pérez, Hai-Long Wu

Notice

First published in London, United Kingdom, 2019 by IntechOpen
IntechOpen is the global imprint of INTECHOPEN LIMITED, registered in England and Wales, registration number: 11086078, The Shard, 25th floor, 32 London Bridge Street
London, SE19SG – United Kingdom
Printed in Croatia

British Library Cataloguing-in-Publication Data
A catalogue record for this book is available from the British Library

Additional hard copies can be obtained from orders@intechopen.com

Chemometrics and Data Analysis in Chromatography, Edited by Vu Dang Hoang
p. cm.
Print ISBN 978-1-78923-835-8
Online ISBN 978-1-78923-836-5

For EU product safety concerns:
IN TECH d.o.o., Prolaz Marije Krucifikse Kozulić 3, 51000 Rijeka, Croatia,
info@intechopen.com or visit our website at intechopen.com.

Meet the editor

Vu Dang Hoang completed his doctorate in pharmaceutics at the University of Strathclyde, UK, in 2005 and conducted postdoctoral research at the Ecole Nationale d'Ingénieurs des Techniques des Industries Agricoles et Alimentaires, France, in 2006. He has been lecturing at the Department of Analytical Chemistry and Toxicology, Hanoi University of Pharmacy, Vietnam, since 2007. He became an associate professor in the field of drug quality control in 2015. His research interests include chemico-physical characterization of topical drug delivery systems and chemometrics-based methods for the analysis of drugs in pharmaceutical dosage forms and biological fluids. So far, he has authored more than 30 publications in pharmaceutical research and analytical chemistry in leading national and international peer-reviewed journals.

Contents

Preface

In 1901, the Russian botanist Mikhail S. Tsvet invented adsorption chromatography when devising a separation technique for plant pigments. The theoretical rationale behind chromatography, which separates components of a mixture by having relative amounts of each solute distributed between a moving fluid stream (i.e. mobile phase) and a stationary phase, was then investigated by Archer J.P. Martin and Richard L.M. Synge. Chromatographic science was hugely developed by the discovery of gas-liquid chromatography by Archer J.P. Martin and Anthony T. James in 1952. In comparison with older techniques (such as crystallization, solvent extraction, and distillation), chromatography offers unequaled resolving power. It can separate all the components of a multicomponent chemical mixture without prior knowledge of all the target substances present in the sample. Some chromatographic techniques can detect substances present at below nanogram levels, thus making it extensively applied in biomedical, pharmaceutical, environmental, and forensic sciences. This book intends to provide the readers with an up-to-date application of chemometrics and data analysis to chromatographic methods.

Assoc. Prof. Dr. Vu Dang Hoang
Department of Analytical Chemistry and Toxicology
Hanoi University of Pharmacy
Hoan Kiem, Hanoi, Vietnam

Chapter 1

Introductory Chapter: Mathematical Methods in Liquid Chromatography - The State-of-the-Art Developments and Challenges

Vu Dang Hoang

Additional information is available at the end of the chapter

http://dx.doi.org/10.5772/intechopen.84589

1. Introduction

Liquid chromatography approaches account for a wide range of analytical applications that really have impact on our daily life, especially for pharmaceutical, biological and food analysis. It is the undeniable fact, even for those not initiated in the world of science [1].

Technically speaking, it is used to separate out a mixture into its individual parts based on the interactions of the sample and the mobile and stationary phases. There are several different types of liquid chromatography, in which the mobile phase is a liquid and the separation can be performed either in a column or on a plane. It is noteworthy that HPLC is the most popular chromatographic technique presently employed. This acronym stands for high-performance (or high-pressure) liquid chromatography, featuring the utilization of pumps to pass a relatively high pressurized liquid solvent through a column packed with very small particles.

Although HPLC is a versatile separation tool, the coupling of liquid chromatographic analysis with mathematical methods has been increasingly proved to be an economical alternative to resolve any problematic situation without using sophisticated instruments in modern laboratories. In this introductory chapter, the state-of-the-art developments and challenges of the application of mathematical methods in liquid chromatography are briefly described in a context to fit nicely with the scope of this book, specifically aimed at the exploitation of chemometrics and data analysis in chromatographic science.

2. Mathematical methods in liquid chromatography

2.1. Method development and validation

It was recognized that method development in HPLC analysis is often a troublesome and bewildering task for the newcomer approaching the problem in a very much "trial and error" manner [2]. This is because chromatographic separation of the analytes is a function of the controllable variables (e.g., mobile phase velocity and composition, column physico-chemical properties, separation temperature, detection wavelength, etc.) and the uncontrolled variables (noise, drift, column performance). Although the separation could be reasonably achieved with little consideration of the fundamental aspects of these variables, it would be optimized with a full consideration of their relevance.

In analytical chemistry, the optimization of an analytical response (called as the dependent variable) is to find the optimal settings or conditions for a number of experimental factors (called the independent variables) by using mathematical terms or rules. As early as 1979, Massart presented this issue at the 27th International Congress of Pure and Applied Chemistry, specifically addressing the methods unknown to the majority of analytical chemists such as information theory, autocovariance curves as applied in continuous analysis, multi-criteria analysis, feature reduction in pattern recognition and operations research [3]. At that time, method optimization was primarily performed via Dantzig's *simplex* algorithm (or *simplex* method) [4]. For HPLC separations, however, there are significant disadvantages associated with simplex optimization because it is generally unable to assess the quality of a located optimum (e.g., the global optimum could not be obtained due to a change in peak elution orders in successive separations) and it requires a relatively large number of experiments to locate optimum separation conditions [5].

In a tutorial, Hayashi and Matsuda described the total optimization of a chromatographic process using analysis criteria such as (i) precision described by the Shannon mutual information (ϕ) and (ii) efficiency denoted by the transmission speed (υ) of the information (ϕ) [6]. For demonstration, the authors showed the optimum condition of the HPLC analysis of an antipyretic mixture, corresponding to the maximal amount of (ϕ) and/or (υ) among all the operating conditions examined, e.g., mobile phase composition and velocity, organic modifier fraction, column length, detection wavelength and amount of internal standard.

To optimize mobile phases in isocratic RP-HPLC analysis, Coenegracht et al. guided the methodology to locate the global optima in terms of separation and analysis time of isoeluotropic and non-isoeluotropic ternary and quaternary eluent mixtures, using regression techniques combined with multi-criteria decision-making [7]. It was suggested that from the quadratic models, different separation criteria such as the resolution or selectivity of the worst separated pair of peaks could be calculated as well as the capacity factor of the last peak could be predicted as a measure of analysis time.

Statistical modeling of the chromatographic process shows that encountering two or more partially overlapped peaks is highly probable in a chromatogram (e.g., see **Figure 1**). In the 1990s, online multiwavelength absorptiometric detection was often used to generate the additional

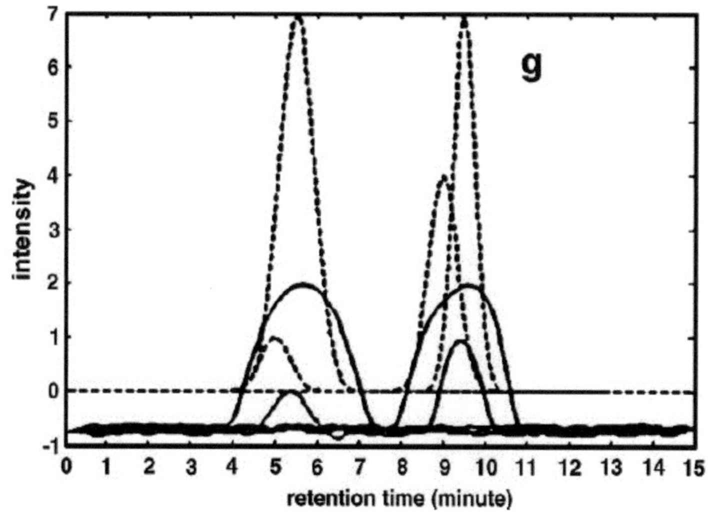

Figure 1. The rank graphs obtained by fixed-size moving window evolving factor analysis (solid line) and pure chromatograms (dotted line) for the simulated four-component data [8].

data required to facilitate peak-purity assessment in LC [9]. Using photodiode array technology, there was still no single algorithm capable of fulfilling all the requirements of peak-purity assessment methods, especially for structurally similar impurities co-eluted with a parent compound. In this case, mass spectrometry (as a reliable and robust LC detector) should be used with multivariate statistical analysis to overcome the problem of spectral discrimination with a higher degree of confidence.

Computerized-assisted optimization (optimization in silico) is an attractive tool (i.e., robust and reliable) for LC method development (e.g., see **Figure 2**). It could be applied for the prediction of analyte retention and optimization of separations in ion chromatography, based on retention time and peak width modeling data. Provided that numerous isocratic and/or gradient steps in ion chromatographic analysis of very complex mixtures involve a large input in time, in silico approaches can offer rapid algorithms for prediction of retention time [11].

In 1992, Carlson [12] drew the attention of chemists on the use of statistical experimental design with a tutorial emphasizing on the steps to be taken before conducting any multivariate design of a screening experiment.

Nowadays, the multivariate optimization techniques have been increasingly employed in the field of analytical chemistry, in particular the surface response methodologies (e.g., see **Figure 3**). For preliminary assessment of experimental factors, the two-level full factorial design was often used, whereas the central composite design was most utilized for determination of critical conditions using quadratic models [14]. With the greater availability of statistical software and overall capacity of modern instrumentation, the popularity of experimental design has not been surprisingly increasing. In chromatographic analysis, it is devoted to showing lack of significant effects in robustness studies for method validation (e.g., Plackett-Burman designs) and identifying significant factors in response to optimization for method

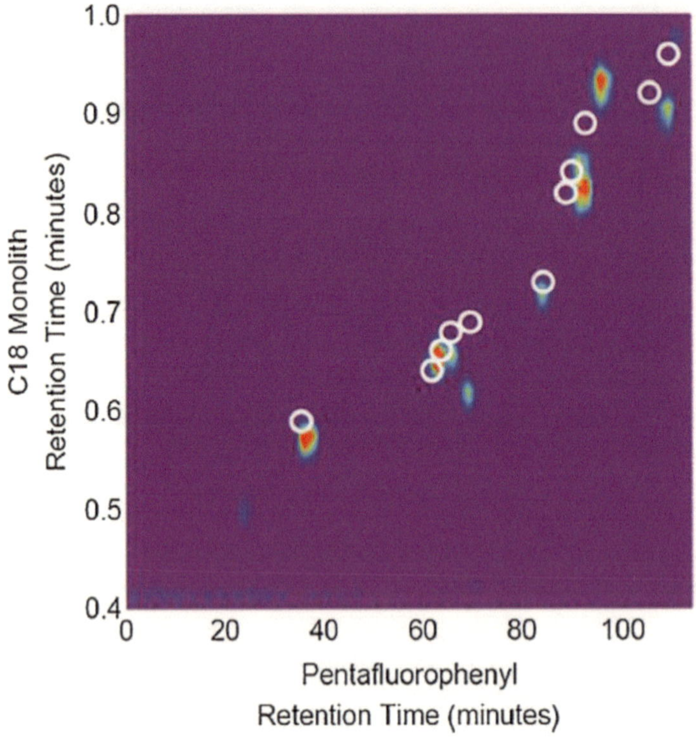

Figure 2. Overlaid DryLab1 and actual two-dimensional separation of the model methamphetamine seizure sample [10].

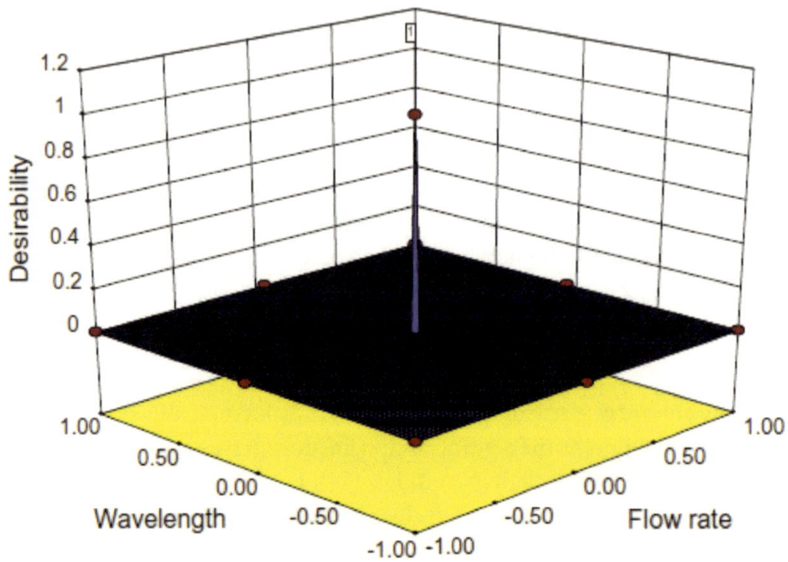

Figure 3. 3D surface response plot of desirability for optimization of HPLC factors: peak area, tailing factor (5%) and number of theoretical plate [13].

development (e.g., fractional factorial designs and their extensions such as central composite designs, Box-Behnken and Doehlert designs). It is noticed that D-optimal designs are now becoming more popular and particularly useful when the factor space is not uniformly accessible (when combinations of solvent composition and solute concentration are not possible) and/or there is a constraint on the total number of experiments that can be done [15]. Doehlert uniform shell designs have been implemented in LC analysis for the development and improvement of sample preparation by exploring up to four factors and the optimization of specific instrumental parameters by exploring up to three factors [16].

Recently, chemometric approaches, i.e., the Quality by Design paradigm for method development and the Six Sigma practice as a quality indicator in chromatography, have been acknowledged as precious complements to HPLC practices [17]. Although Design of Experiment (DOE) application in chromatography gives excellent separation desirability, global satisfaction is still unattainable. The treatment of Quality by Design paradigm, initiated by USFDA and ICH, can locate the global optimal condition of a robust HPLC method with fewer method transfer or failures and issues by establishing a comprehensive design space as a function of DOE philosophy. On the other hand, Six Sigma is a set of statistical techniques and tools used to indicate a process yield in terms of the percentage of defect-free outputs, i.e., in a Six Sigma process, 99.99966% of all outputs are statistically expected to be free of defects. The characteristics of a chromatographic process may be thus amenable to Six Sigma overall process control, e.g., reduction in organic solvent used.

In an effort to improve LC analysis, it is often worth trying to replace a column with either an identical or a totally different one. This could be effectively done, based on the classification of chromatographic stationary phases by a number of approaches, i.e., radar plots, hierarchical cluster analysis, principal component analysis, calculation of a distance factor from a reference column based on Pythagorean Theorem, and two-dimensional diagrams plotting one property versus another [18]. Moreover, chromatographic resolution and analysis time could be further enhanced by modulation of the stationary phase through a serial coupling of columns. Full benefit of such coupling is only achieved through the application of interpretive (i.e., based on model) optimization of columns (nature and length) and solvent (isocratic or gradient mobile phase) [19].

Basically, a chemometric approach can help systematize the knowledge of factors affecting retention in RP-HPLC. By this means, quantitative structure (reversed-phase)–retention relationships (QSRRs) could rationalize RP-HPLC retention mechanism by employing molecular descriptors expected to model fundamental intermolecular interactions and solvatochromic parameters of similar potency for retention prediction [20]. In literature, different modeling methodologies were reported for QSRR approaches for RP-HPLC, e.g., principal component analysis (e.g., see **Figure 4**) and decision trees; these methodologies include artificial neural networks, partial least squares, uninformative variable elimination partial least squares, stochastic gradient boosting for tree-based models, random forests, genetic algorithms, multivariate adaptive regression splines, and two-step multivariate adaptive regression splines [22]. In thin layer chromatography (TLC) analysis, QSRR studies were employed to illustrate the relationship between retention and lipophilicity of solutes [23]. Nonetheless, not all physicochemical descriptors correlate strongly with retention data, and there is no need to display retention data in the form of an equation, given a small number of compounds involved [24].

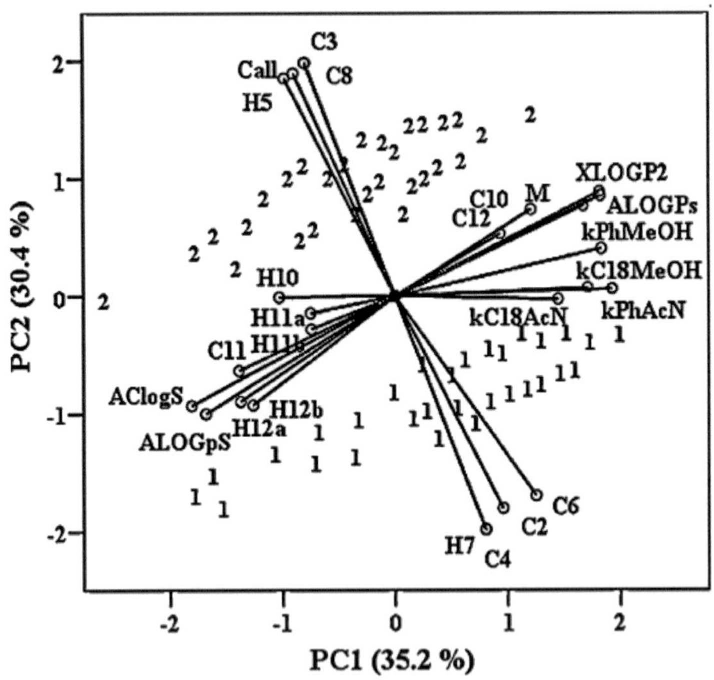

Figure 4. Biplot of principal component analysis for 59 objects (esters of alkoxyphenylcarbamic compounds), four studied retention factors and further 21 significant variables. 1, ortho-derivatives and 2, meta-derivatives [21].

Hydrophilic interaction liquid chromatography (HILIC) is an alternative HPLC mode that utilizes hydro-organic mobile phases with a high organic content and a hydrophilic stationary phase for separating predominantly polar compounds and charges substances in complex matrices. With an enormous growth in the application of HILIC, computer-aided modeling (most importantly the QSRR modeling strategy) proved to be useful in understanding the retention mechanism, classification of stationary phases, prediction of retention times, optimization of chromatographic conditions, and interaction effects of chromatographic factors [25].

In addition to QSRR studies, molecular simulation methods for RP-HPLC modeling (such as molecular dynamics and Monte Carlo) are able to elucidate not only the retention mechanism for different analytes but also (i) the structure and dynamics of the bonded phase and its interface with the mobile phase and (ii) the interactions of analytes with the bonded phase. While the former can provide information on chain dynamics and transport properties, the latter is uniquely suited for the investigation of phase and sorption equilibria underlying RPLC retention [26].

2.2. Data analysis

Despite of being widely applied to the analysis of real-world samples, one-dimensional chromatography cannot always guarantee sufficient resolving power for separation of target compounds. As a result, multidimensional chromatography is proposed for delivering heightened separation performance for complex and difficult substances [27]. A conventional heart-cutting multidimensional chromatographic technique uses multiple columns, each with a different stationary phase, positioned at right angles to each other to selectively choose elements unthoroughly processed from the first column and transfer them to the second column for better separation. In contrast, in a comprehensive multidimensional chromatography, the entire sample is subjected to separation in both dimensions with the following requirements: (i) any two components separated in the first dimension must remain separated in the second dimension and (ii) elution profiles from both dimensions are preserved [28]. In principle, the implementation of two-dimensional chromatography could be realized by combining the chromatographic separations in space (LC^x) and/or in time (LC^t). Although column-based combinations ($LC^t \times LC^t$) are the most actively pursued, the combination $LC^x \times LC^t$ can provide remarkable time-saving because it allows simultaneous second-dimension separations of all the fractions resolved in the first-dimension separations [29]. It was shown that fast, comprehensive two-dimensional liquid chromatography could offer the huge advantage over one-dimensional one with regard to a tremendous potential increase in resolving power. This is because the peak capacity of the comprehensive system approaches the product of the peak capacities of the first- and second-dimension separations when combining highly orthogonal separation mechanisms [30]. This chromatographic technique (particularly coupled with mass spectrometry detection) has been increasingly applied for the analysis of food samples of both relatively low and highly challenging complexities [31–33], complex Chinese herbal medicines (e.g., see **Figure 5**) [35, 36], synthetic polymers and biopolymers [37]. Data processing of comprehensive two-dimensional chromatography is an area of great expansion with reference to data acquisition and handling, peak detection (i.e., two-concept algorithm, inverted watershed algorithm and multiway chemometric methodologies (such as parallel factor analysis model, target finder algorithms and multivariate curve resolution with alternating least squares)), deconvolution of overlapping peaks and data analysis software [38].

Modern instruments can generate a variety of second- and higher-order measurement data that makes multiway analysis a subject of high-level interest for the analytical community [39]. Separation quality of hyphenated chromatography—(multichannel) spectroscopy instruments—could be reliably and essentially evaluated by using chemometric methods for two-way data treatment to determine the number of chemical components, elution sequence and peak purity [40]. Identification and quantification of banned substances or substances with a specified maximum limit could be chromatographically performed by constructing calibration of a systematic analytical procedure with chromatographic n-way data, with commonly used techniques such as n-way partial least squares, multivariate curve resolution and parallel factor analysis [41]. Numerous works on quantitative analysis were reported for second-order data (matrices from unidimensional chromatography with multivariate detection or from two-dimensional chromatography) or third-order data (three-dimensional data arrays from two-dimensional chromatography with multivariate detection), with special attention paid to processing algorithms to cope with the ubiquitous phenomenon of inter-run variation in time shift [42].

By definition, chromatographic fingerprinting is a quality control model that looks at the "complete information" or comprehensiveness of the chromatogram and displays integrated quality information. In recent years, its exploitation has really implied a methodological change on the major paradigm of chromatography. By referring to the common peaks of chromatograms, chromatographic fingerprinting could be applied for identification and authentication of foodstuffs

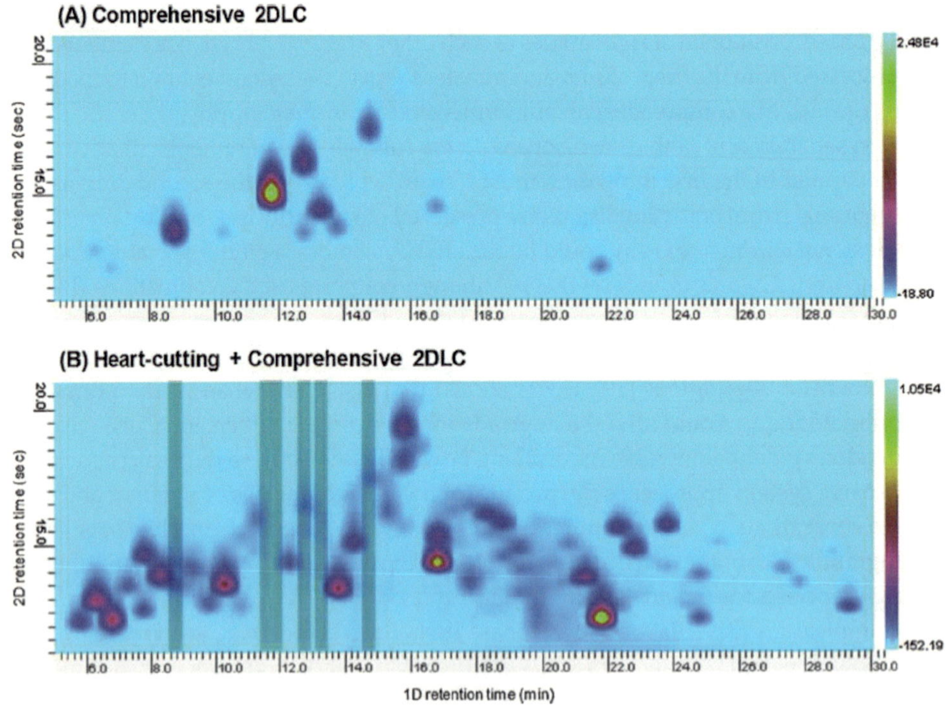

Figure 5. Contour plots for *P. lobata* by using conventional comprehensive 2DLC (A) and heart-cutting + comprehensive 2DLC (B). Dark areas in plot (B) indicate heart-cutting sections. After removing the major compounds by heart-cutting, the minor constituents of the Chinese herbal medicine *Puerariae radix* (Ge-Gen) could be clearly observed on the 2D plot [34].

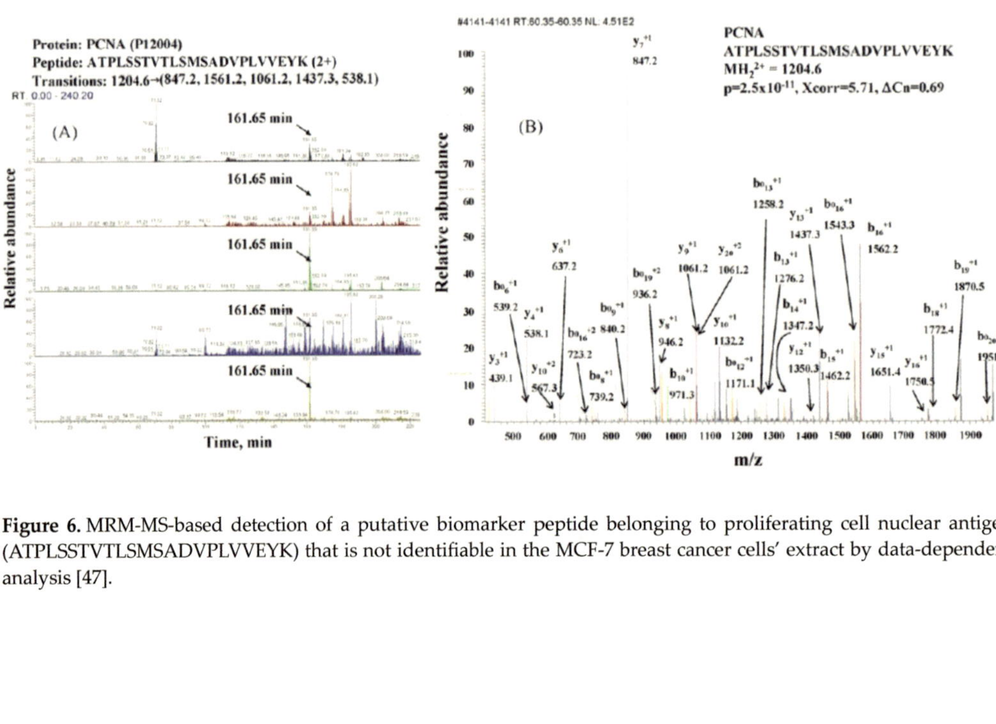

Figure 6. MRM-MS-based detection of a putative biomarker peptide belonging to proliferating cell nuclear antigen (ATPLSSTVTLSMSADVPLVVEYK) that is not identifiable in the MCF-7 breast cancer cells' extract by data-dependent analysis [47].

[43], traditional Chinese medicines [44], and fats and oils [45]. Specific chemometric tools are used in such studies, aiming at (i) verification of the similarity between chromatographic signals, (ii) resolution of overlapped chromatographic signals, and (iii) classification of samples regarding diverse criteria.

Modern combination of liquid chromatography with mass spectrometry, in particular matrix-assisted laser desorption ionization time-of-flight mass spectrometry (MALDI-TOF MS), has quickly established LC–MS as a popular tool used by the proteomics research community to search for new potential markers associated with pathological conditions over the last decade [46]. Many different chemometric approaches currently used for LC–MS spectrum processing in biomarker discovery are mainly for reducing the background noise and properly classifying the samples to the studied groups based on statistically significant features selected by chemometric algorithms (e.g., see **Figure 6**).

3. Conclusion

Based on the above overview, it is reasonable to state that mathematical methods provide effective means for all critical stages in LC analysis with a considerable degree of flexibility and adaptability to different types of practical problems. Nevertheless, despite the fact that they have been continually improved since the 1990s, most chromatographers continue to develop LC procedures manually on the basis of their own experience and chromatographic training. Therefore, it requires new more user-friendly chemometric tools that are better adapted to situation-driven development of chromatographic separation so that computer-assisted method development will become a routine procedure in chromatography laboratories.

Author details

Vu Dang Hoang

Address all correspondence to: hoangvd@hup.edu.vn

Department of Analytical Chemistry and Toxicology, Hanoi University of Pharmacy, Hanoi, Vietnam

References

[1] Uses of Chromatography in Everyday Life Chromatography Today [Internet]. Available from: https://www.chromatographytoday.com/news/industrial-news/39/breaking-news/5-uses-of-chromatography-in-everyday-life/32639 [Accessed: January 21, 2019]

[2] Berridge JC. Chemometrics and method development in high-performance liquid chromatography. Part 1: Introduction. Chemometrics and Intelligent Laboratory Systems. 1988;**3**:175-188

[3] Massart DL. Optimization in Analytical Chemistry: A Multivariate and Multicriteria Problem. In: 27th International Congress of Pure and Applied Chemistry. Plenary and Invited Lectures Presented at the 27th IUPAC Congress, 27-31 August 1979, Helsinki, Finland; 1980. pp. 367-376

[4] Burton KWC, Nickless G. Optimisation via simplex. Part I. Background, definitions and a simple application. Chemometrics and Intelligent Laboratory Systems. 1987;**1**:135-149

[5] Berridge J. Simplex optimization of high-performance liquid chromatographic separations. Journal of Chromatography. 1989;**485**:3-I4

[6] Hayashi Y, Matsuda R. Optimization theory of chromatography. Chemometrics and Intelligent Laboratory Systems. 1993;**18**:1-16

[7] Coenegracht PMJ, Smilde AK, Metting HJ, Doornbos DA. Comparison of optimization methods in reversed-phase high-performance liquid chromatography using mixture designs and multi-criteria decision making. Journal of Chromatography. 1989;**485**:195-217

[8] Lu X, Jiang J-H, Hui C, Xu W-J, Wu H-L, Shen G-L, et al. A criterion for evaluation of chromatographic separation efficiency based on chemometric data treatment: Peak-purity weighted resolution. Chemometrics and Intelligent Laboratory Systems. 2007;**85**(2):195-202

[9] Castledine JB, Fell AF. Strategies for peak-purity assessment in liquid chromatography. Journal of Pharmaceutical and Biomedical Analysis. 1993;**11**(1):1-13

[10] Andrighetto LM, Burns NK, Stevenson PG, Pearson JR, Henderson LC, Bowen CJ, et al. In-silico optimisation of two-dimensional high performance liquid chromatography for the determination of Australian methamphetamine seizure samples. Forensic Science International. 2016;**266**:511-516

[11] Ng BK, Tan TTY, Shellie RA, Dicinoski GW, Haddad PR. Computer-assisted simulation and optimisation of retention in ion chromatography. TrAC, Trends in Analytical Chemistry. 2016;**80**:625-635

[12] Carlson R. Preludes to a screening experiment. A tutorial. Chemometrics and Intelligent Laboratory Systems. 1992;**14**:103-114

[13] Kumar L, Sreenivasa Reddy M, Managuli RS, Pai KG. Full factorial design for optimization, development and validation of HPLC method to determine valsartan in nanoparticles. Saudi Pharmaceutical Journal. 2015;**23**(5):549-555

[14] Ferreira SLC, Lemos VA, de Carvalho VS, da Silva EGP, Queiroz AFS, Felix CSA, et al. Multivariate optimization techniques in analytical chemistry—An overview. Microchemical Journal. 2018;**140**:176-182

[15] Hibbert DB. Experimental design in chromatography: A tutorial review. Journal of Chromatography. B, Analytical Technologies in the Biomedical and Life Sciences. 2012; **910**:2-13

[16] Araujo P, Janagap S. Doehlert uniform shell designs and chromatography. Journal of Chromatography. B, Analytical Technologies in the Biomedical and Life Sciences. 2012; **910**:14-21

[17] Sahu PK, Ramisetti NR, Cecchi T, Swain S, Patro CS, Panda J. An overview of experimental designs in HPLC method development and validation. Journal of Pharmaceutical and Biomedical Analysis. 2018;**147**:590-611

[18] Lesellier E, West C. Description and comparison of chromatographic tests and chemometric methods for packed column classification. Journal of Chromatography. A. 2007; **1158**(1-2):329-360

[19] Alvarez-Segura T, Torres-Lapasio JR, Ortiz-Bolsico C, Garcia-Alvarez-Coque MC. Stationary phase modulation in liquid chromatography through the serial coupling of columns: A review. Analytica Chimica Acta. 2016;**923**:1-23

[20] Kaliszan R. Quantitative structure-retention relationships applied to reversed-phase high-performance liquid chromatography. Journal of Chromatography. A. 1993;**656**: 417-435

[21] Durcekova T, Boronova K, Mocak J, Lehotay J, Cizmarik J. QSRR models for potential local anaesthetic drugs using high performance liquid chromatography. Journal of Pharmaceutical and Biomedical Analysis. 2012;**59**:209-216

[22] Put R, Vander Heyden Y. Review on modelling aspects in reversed-phase liquid chromatographic quantitative structure-retention relationships. Analytica Chimica Acta. 2007;**602**(2):164-172

[23] Ciura K, Dziomba S, Nowakowska J, Markuszewski MJ. Thin layer chromatography in drug discovery process. Journal of Chromatography. A. 2017;**1520**:9-22

[24] Heberger K. Quantitative structure-(chromatographic) retention relationships. Journal of Chromatography. A. 2007;**1158**(1-2):273-305

[25] Taraji M, Haddad PR, Amos RIJ, Talebi M, Szucs R, Dolan JW, et al. Chemometric-assisted method development in hydrophilic interaction liquid chromatography: A review. Analytica Chimica Acta. 2018;**1000**:20-40

[26] Lindsey RK, Rafferty JL, Eggimann BL, Siepmann JI, Schure MR. Molecular simulation studies of reversed-phase liquid chromatography. Journal of Chromatography. A. 2013;**1287**:60-82

[27] Dugo P, Cacciola F, Kumm T, Dugo G, Mondello L. Comprehensive multidimensional liquid chromatography: Theory and applications. Journal of Chromatography. A. 2008; **1184**(1-2):353-368

[28] Francois I, Sandra K, Sandra P. Comprehensive liquid chromatography: Fundamental aspects and practical considerations—A review. Analytica Chimica Acta. 2009;**641**(1-2):14-31

[29] Guiochon G, Marchetti N, Mriziq K, Shalliker RA. Implementations of two-dimensional liquid chromatography. Journal of Chromatography. A. 2008;**1189**(1-2):109-168

[30] Stoll DR, Li X, Wang X, Carr PW, Porter SE, Rutan SC. Fast, comprehensive two-dimensional liquid chromatography. Journal of Chromatography. A. 2007;**1168**(1-2):3-43

[31] Herrero M, Ibanez E, Cifuentes A, Bernal J. Multidimensional chromatography in food analysis. Journal of Chromatography. A. 2009;**1216**(43):7110-7129

[32] Tranchida PQ, Donato P, Cacciola F, Beccaria M, Dugo P, Mondello L. Potential of comprehensive chromatography in food analysis. TrAC, Trends in Analytical Chemistry. 2013; **52**:186-205

[33] Cacciola F, Dugo P, Mondello L. Multidimensional liquid chromatography in food analysis. TrAC, Trends in Analytical Chemistry. 2017;**96**:116-123

[34] Qiao X, Song W, Ji S, Li YJ, Wang Y, Li R, et al. Separation and detection of minor constituents in herbal medicines using a combination of heart-cutting and comprehensive two-dimensional liquid chromatography. Journal of Chromatography. A. 2014;**1362**:157-167

[35] Cao JL, Wei JC, Chen MW, Su HX, Wan JB, Wang YT, et al. Application of two-dimensional chromatography in the analysis of Chinese herbal medicines. Journal of Chromatography. A. 2014;**1371**:1-14

[36] Ji S, Wang S, Xu H, Su Z, Tang D, Qiao X, et al. The application of on-line two-dimensional liquid chromatography (2DLC) in the chemical analysis of herbal medicines. Journal of Pharmaceutical and Biomedical Analysis. 2018;**160**:301-313

[37] Baumgaertel A, Altuntas E, Schubert US. Recent developments in the detailed characterization of polymers by multidimensional chromatography. Journal of Chromatography. A. 2012;**1240**:1-20

[38] Matos JT, Duarte RM, Duarte AC. Trends in data processing of comprehensive two-dimensional chromatography: State of the art. Journal of Chromatography. B, Analytical Technologies in the Biomedical and Life Sciences. 2012;**910**:31-45

[39] Escandar GM, Goicoechea HC, Munoz de la Pena A, Olivieri AC. Second- and higher-order data generation and calibration: A tutorial. Analytica Chimica Acta 2014;**806**:8-26

[40] Xu L, Tang LJ, Cai CB, Wu HL, Shen GL, Yu RQ, et al. Chemometric methods for evaluation of chromatographic separation quality from two-way data—A review. Analytica Chimica Acta. 2008;**613**(2):121-134

[41] Ortiz MC, Sarabia L. Quantitative determination in chromatographic analysis based on n-way calibration strategies. Journal of Chromatography. A. 2007;**1158**(1-2):94-110

[42] Arancibia JA, Damiani PC, Escandar GM, Ibanez GA, Olivieri AC. A review on second- and third-order multivariate calibration applied to chromatographic data. Journal

of Chromatography. B, Analytical Technologies in the Biomedical and Life Sciences. 2012;**910**:22-30

[43] Cuadros-Rodriguez L, Ruiz-Samblas C, Valverde-Som L, Perez-Castano E, Gonzalez-Casado A. Chromatographic fingerprinting: An innovative approach for food 'identitation' and food authentication—A tutorial. Analytica Chimica Acta. 2016;**909**:9-23

[44] Liang XM, Jin Y, Wang YP, Jin GW, Fu Q, Xiao YS. Qualitative and quantitative analysis in quality control of traditional Chinese medicines. Journal of Chromatography. A. 2009;**1216**(11):2033-2044

[45] Bosque-Sendra JM, Cuadros-Rodriguez L, Ruiz-Samblas C, de la Mata AP. Combining chromatography and chemometrics for the characterization and authentication of fats and oils from triacylglycerol compositional data—A review. Analytica Chimica Acta. 2012;**724**:1-11

[46] Hajduk J, Matysiak J, Kokot ZJ. Challenges in biomarker discovery with MALDI-TOF MS. Clinica Chimica Acta. 2016;**458**:84-98

[47] Armenta JM, Perez M, Yang X, Shapiro D, Reed D, Tuli L, et al. Fast proteomic protocol for biomarker fingerprinting in cancerous cells. Journal of Chromatography. A. 2010;**1217**(17):2862-2870

Chapter 2

Chemometrics-Based TLC and GC-MS for Small Molecule Analysis: A Practical Guide

Juan Vázquez-Martínez and Mercedes G. López

Additional information is available at the end of the chapter

http://dx.doi.org/10.5772/intechopen.77160

Abstract

Nowadays, thin-layer chromatography (TLC) and gas chromatography/mass spectrometry (GC-MS) instruments can produce more data than even before. At this point, the use of mathematical and statistical tools has provided the key to resolve the information overload. In this chapter, a practical guide is provided for the TLC and GC-MS analysis of short-chain fatty acids (SCFAs), amino acids, and monosaccharides. A methodology for extracting and transforming the chromatographic data to a suitable format for chemometrics is described. Furthermore, a procedure for chemometric analysis based on principal components analysis and clustering analysis is suggested.

Keywords: thin layer chromatography, gas chromatography coupled to mass spectrometry, amino acids, short-chain fatty acids, monosaccharides, principal components analysis, clustering analysis

1. Introduction

Chemometrics can be defined as the application of mathematical and/or statistical methods to chemical analysis [1]. Like any aspect of mathematics, it requires the use of numbers obtained from measured values of chemical variables [1, 2]. There exist a large number of instrumental techniques for analytical chemistry, which have certain advantages for specific metabolites or matrices. *Grosso modo*, we can mention spectroscopic (infrared, ultraviolet-visible, X-ray, etc.), mass spectrometric, chromatographic, electrochemical and thermal methods, and hybrid techniques [3, 4]. The advances in electronics and computer-assisted data processing have provided powerful instruments that obtain more information than can be analyzed by using traditional data analysis methods. Thus, at this point, chemometrics appears as the preferred choice for the analysis of these complex data [1, 5].

Chromatography comprises the separation techniques based on the partition of the analytes between a mobile phase and a stationary phase. It is extensively used in food industry, pharmaceutical sciences, and natural products science and technology [3, 4] due to its high sensitivity, selectivity, and reproducibility. The versatility of chromatographic techniques allows analyzing a wide range of metabolites, from low-molecular-weight aliphatic gases to complex high-molecular-weight polymeric substances.

In this chapter, we will describe the state-of-the-art in the use of chemometrics-based thin-layer chromatography (TLC) and gas chromatography coupled to mass spectrometry (GC-MS), paying special attention to the analysis of low-molecular-weight metabolites such as amino acids, short-chain fatty acids (SCFAs), and monosaccharides (referred herein as small molecules). In brief, schematic representations of the advantages and some characteristics of TLC and GC-MS for small molecule analysis are shown in **Figure 1**.

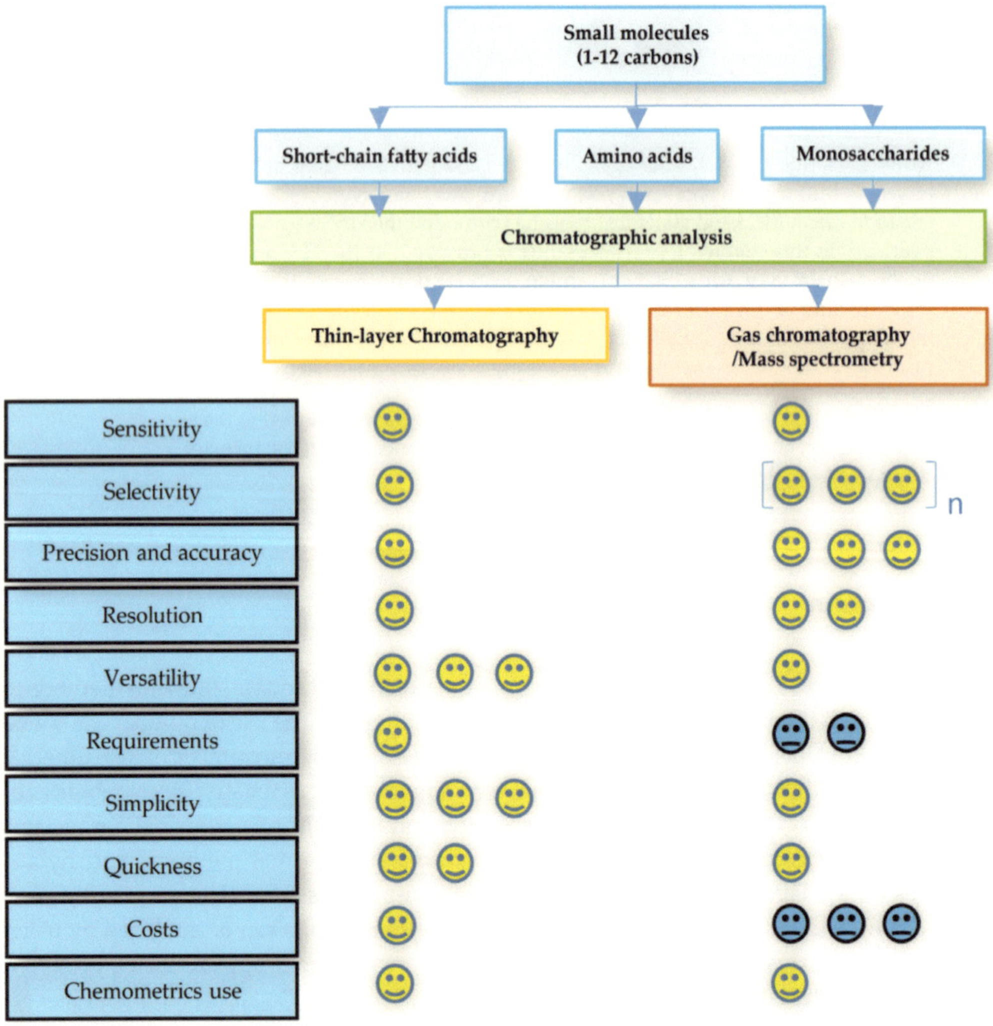

Figure 1. Comparison of some characteristics between TLC and GC-MS for small molecule analysis.

2. TLC analysis of small molecules

2.1. Generalities

TLC is a planar chromatographic technique extensively used due to its rapidity, versatility, and affordable equipment [6]. Bit by bit, it has been relegated as a chromatographic technique for screening tests; however, TLC analysis of some metabolites may represent the best option. The versatility of TLC results from a wide variety of stationary phases used for separation. The common stationary phases include silica with different physical (e.g., pore diameter) and chemical traits (e.g., silanized silica, C_n alkyl-bonded silica) as well as other sorbents such as celluloses, aluminas, and polyamides [6–9]. On the other hand, the improvement of separation of specific metabolites could be largely done by varying the mobile phases from pure nonpolar solvents to complex mixtures of solvents with different polarities [7–9].

The principal measure that can be obtained from a TLC analysis is the retention factor (*Rf*). This parameter is used to describe the migration of components over a TLC plate and is defined as the ratio of the distance traveled by the center of a spot to the distance traveled by the solvent front, both distances are measured from the starting point [7]. By definition,

$$Rf = \frac{\text{component traveled distance}}{\text{mobile phase front traveled distance}} \tag{1}$$

Rf values are always in the range from 0 to 1, or from 0 to 100 if multiplied by 100 (*hRf*) to avoid the decimal point. This parameter is used for component identification, since every compound has a specific *Rf* value for every specific mobile phase. If other compounds comigrate and appear with the same or similar *Rf* value, it is preferable to improve the separation by changing the mobile phase or using a specific visualization reagent [6–8]. Although *Rf* value is very useful for identifying components in TLC and in some cases gathers enough information to qualify and semiquantify a sample [7, 10], it is inadequate for chemometrics-based analysis. For a more precise TLC data handling, it is necessary to complement *Rf* data with numerical values that correspond to the abundance of retained components [10].

An elementary and simple approach for quantitative purpose is the visual comparison of the spot/band intensity of a known sample aliquot with the intensities of a concentration series of known standards, all developed in the same TLC plate. This approach offers semiquantitative results, with precision (expressed as absolute deviation) and accuracy (expressed as experimental error) ranged from 10 to 30% [6–9]. Concentration estimation via visual comparison depends on the interpretation of the analyst. To reduce the interanalyst interpretation, the standard concentration series should be very close to the sample aliquot intensity. Other simple approaches for quantitative TLC determination include measuring the area of the spot or band directly by approximating the spot/band to a regular geometric figure. However, there is still an error range dependent on the precision of the approximation with these approaches.

2.2. TLC method development

SCFAs are fatty acids with 2–12 carbon atoms, although other authors differentiate SCFAs with 2–6 carbon atoms [11] from medium-chain fatty acids with 7–12 carbon atoms [11, 12].

These metabolites are produced by dietary fiber fermentation in the colon [13]. Some SCFAs, like butyric acid, have medical relevance since they contribute to colon health by working as anti-inflammatory and energy-source compounds [11–13]. For TLC analysis of SCFAs, it is necessary to consider the autoesterification and autopolymerization reactions of some organic acids because of their dehydration from a dilute aqueous solution, particularly for lactic acid [14]. These reactions are evidenced by the presence of two or more spots in a chromatogram of the respective acid.

The term "amino acids" is used to group the organic compounds that contain amine and carboxyl functional groups, along with side chain group. Some of these compounds are the building units of proteins. Commonly, this term refers only to the 20 amino acids of the genetic code, but there exist about 500 naturally occurring compounds that are chemically amino acids [15, 16]. The TLC techniques to analyze this group of compounds exploit the physicochemical characteristics of the amino and carboxylic moieties, although the separation on a TLC plate is due to the characteristics of the side chain [17–19]. It is noteworthy that two-dimensional TLC development should be required to better separate a complex mixture of more than 10 amino acids [19–21].

Monosaccharides belong to a large family of natural products (i.e., carbohydrates) with the general formula $C_n(H_2O)_n$, the basic structures consisting of five carbons (pentoses) or six carbons (hexoses). They are either polyhydroxyaldehydes or polyhydroxyketones. The *alpha* hydroxyl group of some monosaccharides can be replaced by another substituent such as hydrogen in deoxy sugars and amino group in amino sugars. Furthermore, they can be oxidized to acidic sugars or reduced to polyols [8, 22]. Although they mainly exist as their cyclic hemiacetals or hemiketals, it is necessary to consider the equilibrium of both cyclic and acyclic forms for appropriate chromatographic analysis [8]. Monosaccharides are the most polar compounds of the small molecules here mentioned, and the TLC techniques to analyze this group exploit this feature.

Table 1 summarizes the condition and procedure for TLC analysis of abovementioned small molecules. Because most of these molecules are colorless and nonfluorescent under ultraviolet and visible light, the use of a derivatization reagent is a must for their visualization.

2.3. TLC data treatment

Direct optical quantification in TLC can be realized by using a slit-scanning densitometer. In this technique, the absorbance or emitted fluorescence of the components separated in a TLC plate is measured. According to the compound nature or their derivatives' spectral characteristics, halogen or tungsten (for the visible range) and deuterium lamps (for the UV region) can be used; nevertheless, better results are usually obtained with absorption of UV light on regular layers. On the other hand, on layers with incorporated phosphor, the compound abundance can be quantified upon UV absorption by dark zones on a fluorescent background (fluorescence quenching) [7]. This type of equipment can be computer-controlled, performing automated and accurate data acquisition and processing. At present, commercially available scanning TLC densitometers have common technical characteristics such as spectral range

Technical parameters	SCFAs	Amino acids	Monosaccharides
Stationary phase	20 × 20 cm aluminum-backed plates precoated with silica gel 60 (0.2 mm layer) [14]	20 × 20 cm silica gel 60 plates [7, 18–21]	(1) 10 × 20 cm high-performance thin-layer cellulose plates [8, 25] (2) 10 × 20 cm aluminum-backed plates precoated with silica gel 60 [26, 27]
Mobile phase	(1) *n*-Hexane-acetone (4:1, *v*/*v*) for C6 to C8 SCFAs [14] (2) Acetone-water-chloroform-ethanol-aqueous ammonia (30:1:3:5: 11, *v*/*v*) for C2 to C5 SCFAs and lactic acid [14, 23]	Two-dimensional development: (i) *n*-butanol-acetic acid-water (4:1:1, *v*/*v*), and (ii) phenol-water (75:25, *w*/*w*) [7, 18–21]	For stationary phase 1: ethyl acetate-pyridine-acetic acid-water (A, composed of 35:15:1:7, *v*/*v* and B, composed of 35:15:1:9, *v*/*v*) [25] For stationary phase 2: C, composed of *n*-propanol-*n*-butanol-water (12:3:4, *v*/*v*) and D, composed of ethyl acetate-*n*-propanol-acetic acid-water (4:2:2:1, *v*/*v*) [25]
Chamber and TLC plate's postseparation treatment	Saturate chamber 30 min [14] After development, dry TLC plate at room temperature for 24 h or until a complete mobile phase evaporation	Saturate chamber 20 min [18–21] After each development, dry TLC plate at 100°C	For stationary phase 1: Saturate 20 min before each respective development After development, dry TLC plate at 105°C for 5 min [8, 25, 27] For stationary phase 2: Saturate chamber 10 min before each respective development. A two- or three-time development recommended at a proportional distance in TLC plate [26, 28] Dry TLC plate at 105°C for 5 min [28]
Derivatization agent and visualization	0.25 g of methyl red and 0.25 g of bromophenol blue in 100 mL of 70% methanol Derivatized SCFAs (from two to eight carbons) appeared as blue spots, lactic acid as pink-red spot [14, 23]	0.25% ninhydrin in acetone [7, 18] Optional double derivatization: 0.20% picric acid acetone and 0.25% ninhydrin acetone [24] Heat TLC plate at 110°C for 10 min	For stationary phase 1: 1.6 g of *O*-phthalic acid in 100 mL of water-saturated *n*-butanol, containing 0.9 mL of aniline [25] For stationary phase 2: (i) 250 mg of orcinol dissolved in 95 mL of ethanol and 5 mL of sulfuric acid; or (ii) 4 g of diphenylamine, 4 mL of aniline, and 20 mL of 85% phosphoric acid in 180 mL of acetone Afterward, dry TLC plate at room temperature and heat it at 85–100°C for 5–15 min [26, 28]
Limit of detection (LOD)*	8 μg [14, 23]	0.05–0.1 μg [18] 0.04–0.08 μg [24]	For aniline-*O*-phthalic acid reagent: 10 μg [8] For aniline-diphenylamine-phosphoric acid reagent: 10 ng (fluorescence) [8]

*LOQ is calculated by multiplying LOD by 3.333 [29].

Table 1. Technical parameters for TLC analysis of small molecules.

(190–900 nm), data step resolution (25–200 μm), and scanning speed (1–100 mm/s) that translate in a more representative, accurate, and precise data matrix. The principal disadvantage of these densitometers could be the cost of the instrument itself (30,000–60,000 USD) and of its spare parts [30].

A TLC densitometer chromatogram consists of one axis representing the *Rf* values and the other representing the measured absorbance or fluorescence, which can be extracted as a two-column matrix for posterior chemometric analysis. A well-resolved component is characterized by a well-shaped (taller-than-wide) and normally distributed peak in the chromatogram, typically leading to RSD values in the range of 0.5–3% in quantitative high-performance TLC, using the peak area corresponding directly to the compound concentration [7, 10].

Apart from slit-scanning densitometry, a compound can be quantified based on the analysis of its spot/band image obtained from a TLC plate. The instrument used for this purpose is known as video scanner densitometer, and it is certainly coupled to a computer system. For chromatogram data acquisition, image densitometers obtain a picture of the TLC and subsequently measure the color brightness of visible spots. Commonly, the image is obtained under white light and/or UV light (short-wave and/or long-wave radiation) [30]. Since the first step for this technique is the obtaining of a good-resolution colorful image, homemade instruments equipped with a high-quality digital camera can be adapted for image acquisition.

The disadvantages of image densitometers are related to their lower sensitivity and chromatographic resolution than slit-scanner densitometers. As a result, the data matrix loses a lot of information of the components that cannot be detected under a few wavelengths of light applied and may contain more background interference. However, due to their simplicity and lower cost, these instruments are the most popular for densitometric evaluation nowadays [7, 30].

3. GC-MS analysis of small molecules

3.1. Generalities

Basically, GC is a more sophisticated technique than TLC; for a better understanding, some references are suggested [3, 4, 31]. This is the method of choice for separation and detection of permanent inorganic gases and volatile organic compounds in a mixture. It is based on the partitioning of vaporized or gaseous compounds between an inert gas mobile phase and a stationary liquid or solid phase. Helium is the most commonly used carrier gas, but others such as nitrogen or hydrogen can be used. The separation column is packed with a finely divided solid or coated with a thin film of liquid (typically <1 μm). In the market, there is a wide variety of capillary columns, which can be grouped by the polarity of their stationary phases, for achieving better separation and resolution [31].

The affinity of the components of a mixture for the stationary phase depends on their physicochemical characteristics and it impacts directly on their separation and resolution.

Components' separation is based on the "like-dissolves-like" rule that explains the different interaction strengths between the compounds and stationary phase. A stronger compound-stationary phase interaction provokes a longer compound-stationary phase contact, and more time is needed for compounds' migration through the column. This migration time is known as retention time *Rt* (commonly expressed in minutes) and represents one of the principal numeric values that can be obtained by GC for compound characterization. The other value also obtained by GC is the relative abundance of the components that corresponds with the height or area of the respective chromatogram peak. The unit of the relative abundance depends on the type of detector coupled to GC [4, 31].

GC-MS is a hybrid technique, in which a gas chromatograph is coupled to a mass spectrometer via an interface (i.e., a heated metal tube equipped with a temperature controller, connecting the column exit in the gas chromatograph and the entrance to the ion source of the mass spectrometer). GC alone can separate volatile compounds with great resolution, but it cannot identify them properly [4]. MS uses the difference in mass-to-charge ratio (*m*/*z*) of ionized atoms or molecules, providing structural information by identification of distinctive fragmentation patters. Thus, after separation in the GC column, analytical species are transported to the mass spectrometer to be ionized for subsequent mass filtration and detection. GC-MS is a potent tool for modern analytical chemistry that allows separating the compounds in complex mixtures and identifying them effectively with some considerations [3, 31].

Theoretically, MS is based on the analysis of ions moving through a vacuum, and a mass spectrometer must include the ion source (electron or chemical ionization), ion analyzer (quadrupole, ion trap, or time-of-flight), and ion detector [3, 31]. For GC-electron impact-quadrupole-mass spectrometry (GC-EIMS), immediately after the compounds leave the capillary column, they are bombed by an electron beam and fragmented in ions that correspond to molecule fractions. Then, these ions are separated in the quadrupole according to their mass-to-charge ratio (*m*/*z*) before being further sensed and quantified by the detector.

A GC-EIMS chromatogram (named total ion chromatogram or TIC) is a graph showing the relationship between the retention time (*x*-axis) and ion abundance or total ion current data (*y*-axis). Since the TIC comes from the convolution of the individual abundances of all the monitored ions, it can be deconvoluted to obtain each ion distribution. It can be also displayed in three dimensions simultaneously (i.e., a 3D chromatogram recording the number of ions created along with their masses over time). By examining TIC and "slicing" along the third dimension (*m*/*z*) of a chosen peak, the mass spectrum can be evaluated at a given time. When all of these ions come from the same compound, the ion distribution can be plotted in an abundance *versus m*/*z* graph (aka the compound mass spectrum or fragmentation pattern). In the case that mass spectra are obtained under standard conditions, the distribution of ions is always the same independently of GC step and/or MS instrument. This important feature has enabled constructing big mass spectra libraries that serve for precise compound identification [31].

The first step to analyze small molecules or any metabolite by GC-EIMS is to make sure that all analytes can be volatilized under the injection port and oven temperatures. It is known that the vaporization of a specific compound occurs at a given temperature and pressure that, at the time, depends on the number of carbons and the polarity of the compounds (among other properties). Amino acids, SCFAs, and monosaccharides are polar compounds due to their amino, hydroxyl, carbonyl, and carboxylic groups. These functional groups allow intramolecular interactions such as hydrogen and van der Waals bonds, increasing the compounds' boiling point and making their analysis by GC-EIMS difficult. Thus, only some of these compounds can be GC-EIMS analyzed directly, and for the remaining ones, a derivatization step must be required for better analysis.

3.2. GC-MS method development

3.2.1. Direct GC-MS analysis of small molecules

Among small molecules (i.e., amino acids, monosaccharides, and SCFAs), SCFAs are suitable for direct GC-EIMS analysis due to their small aliphatic acyl chain and relatively low vapor pressure [32]. The critical issue in analyzing SCFAs directly is the correct stationary phase selection. Excellent results could be obtained by using a polar-phase capillary column such as Nukol (acid-modified poly-ethylene-glycol phase, Supelco, Sigma-Aldrich). Since these compounds do not have more than 12 carbons, neither a large column nor a long chromatographic method is required. For direct GC-EIMS analysis of SCFA, the chromatographic conditions are as follows: injection port temperature set at 250°C; GC oven temperature initially set at 90°C for 3 min, then subjected to a three-step program [(i) increased to 150°C with 15°C/min ramp rate, (ii) increased to 170°C with 5°C/min ramp rate, and (iii) increased to 200°C with 20°C/min ramp rate and hold for 10 min]; transfer line temperature set at 250°C; stationary phase: a 30 m × 320 μm × 0.25 μm Nukol capillary column (Supelco, Sigma-Aldrich); carrier gas: a constant helium flow of 1 mL/min; standard MS parameters applied (electron energy of 70 eV, ion source temperature set at 230°C, quadrupole analyzer temperature set at 150°C). This method takes around 16 min for analysis of C2 to C12 SCFAs, with measurements obtained at a 25–300 *m/z* range and approximately 3 scans per second.

3.2.2. Derivatization of small molecules for GC-EIMS analysis

The goal of derivatization before GC-MS analysis is to obtain chemical derivatives being more volatile and less reactive than the interest compounds, thus presenting improved chromatographic characteristics [29, 33].

The labile hydrogens of amino acids, SCFAs, and monosaccharides are commonly the target of derivatization procedures. In practice, these hydrogens of the carboxyl, amino, and hydroxyl groups of the abovementioned small molecules can be substituted by trimethylsilyl groups [33] according to the following general derivatization.

First, it is needed to prepare a completely solvent-free sample since water and any solvent (as trace impurities) with labile hydrogens can interfere with the derivatization. To each 5–50 mg of sample, 20 μL of pyridine and 80 μL of *N,O*-bis(trimethylsilyl)trifluoroacetamide supplied with 1% of trimethylchlorosilane (BSTFA + TMCS) are added. For pure standards, a final concentration of 0.5 μg/μL is prepared. Then mixed and heated at 85°C for 15–30 min. After that, pyridine and BSTFA are evaporated under nitrogen flux of the sample; a known volume of isooctane (ca. 200 μL) is subsequently added and 1 μL of this resultant solution can be injected into the GC-MS system in split or splitless mode.

The following GC-EIMS conditions could enable the individual and simultaneous analysis of amino acids, SCFAs, and monosaccharides. The injector temperature is set at 260°C. An HP-5-MS capillary column (30 m × 25 μm × 0.25 μm) is used with helium as carrier gas at a constant flow rate of 1 mL/min. The oven program begins at 45°C (hold for 5 min), then increases at a rate of 10°C/min until 300°C (hold for 25 min). The transfer line temperature is set at 280°C. The mass spectrometer operates at 70 eV of electron energy; the quadrupole and ion-source temperatures are set at 150 and 230°C, respectively. The scan mode is used in the range 40–550 *m*/*z*. Using this method, SCFAs, amino acids, and monosaccharides can be chromatographically separated if their mixture contains appropriate amounts. SCFAs elute from the GC column during the first 10–15 min of analysis, followed by amino acids and monosaccharides later. Due to the interference of solvent and derivatizing reagent traces, this method is not suitable for the analysis of compounds containing less than three to four carbons, for example, for C2-C3 SCFAs, a direct analysis is preferable.

For amino acids, it is important to consider that in the equilibrium, different forms of derivatization products can be found, i.e., the totally derivatized compound (i.e., all the labile hydrogens are derivatized) and partially derivatized compound (i.e., conserving one or more labile hydrogens). If the same conditions are applied and the equilibrium is reached in all samples, the proportion of totally and partially derivatized amino acids remains constant and they are both commonly included in the mass spectra libraries.

A critical point for the GC-EIMS analysis of monosaccharide silyl derivatives is the presence of natural isomers. D-Hexoses naturally have five isomers: two furanoses, two pyranoses, and one linear structure, and their silyl derivatives possess the same isomers too. These derivatives can be separated by GC and the mass spectra of pyranoses and furanoses can be differentiated [34]. In a comparable way to amino acids, in the equilibrium, the proportion of each isomer remains constant. It is important to note that in TLC analysis, the problem of isomers generation after derivatization process is avoided.

For better GC-EIMS results, the use of internal standards is advisable, for example, synthetic methylated SCFAs, synthetic or nonprotein amino acids, and nonbiological polyols or glycosides, as well as carbon- and hydrogen-labeled compounds [32, 35, 36].

3.3. GC-MS data treatment

Nowadays, the extraction of GC-MS data is a simple and fast task, thanks to convenience and availability of the commercial and free software products. For GC-MS data analysis of small molecules, ChemStation Data Analysis and MassHunter software products (Agilent Technologies, Inc.) are commonly used [29].

Using these software packages, TIC data could be extracted to CSV format, that is, a two-column data matrix is obtained, in which are stored the retention time and ion relative abundance values.

Since TIC data contain the distribution of all measured ions, the data of each individual ion can be extracted and merged to obtain a consensus chromatogram that includes the peaks, in which are present all the selected ions. This feature is very useful when characteristic ions have been detected for the compound(s) of interest. For example, for silylated monosaccharides, the 204 and 217 *m/z* ions can be used as a diagnostic tool and by monitoring both ions a complex chromatogram can be simplified to analyze only monosaccharides.

4. Chemometrics-based chromatographic data analysis

At this point, it can be recognized that the data obtained by TLC and GC-EIMS are in the same format, that is, a numeric matrix for subsequent chemometric analysis.

The following paragraphs are considered a practical guide as more than a theoretical description of the chemometric concepts that can be widely reviewed in Refs. [1, 2, 5, 37]. It describes a workflow for obtaining a chemometric analysis of the TLC and GC-MS data (**Figure 2**), which can be divided into three distinct stages: data preprocessing, data processing, and model validation.

4.1. Data preprocessing

This stage includes the baseline correction, retention time or retention factor correction, and noise reduction [38]. It can be applied before or after numeric data extraction. For a prior data extraction, the software products ImageJ, winCATC, or VisionCATS can be used for TLC analysis, whereas OpenChrom, ChemStation Data Analysis, or MassHunter Workstation can be used for GC-MS analysis. For postdata extraction, the data matrices can be edited with statistical software products such as R or MATLAB. Normalization of data is recommended to minimize systemic variation in the data due to changes in instrumental response; this can be done by using internal standards and a subsequent data correction. A data cleanup step can be added to remove artifact peaks or peaks with low repeatability; this can be done by deleting suspected peaks. After doing this, the data must be presented in a single table all under the same conditions and sifted through the same filters.

4.2. Data processing and model validation

Data can be processed by unsupervised [principal component analysis (PCA) and clustering analysis] and supervised multivariate statistical methods [partial least square discriminant analysis (PLS-DA) and between-group analysis (BGA)] [39].

Regarding unsupervised methods, chromatographic data can be analyzed by the visualization of grouping trends and inspection of atypical values using PCA [38, 40]. A way to do a good PCA analysis is to utilize the FactoMineR package [41] in the R software, using R studio as an interface [42, 43] and considering eigenvalues to obtain suitable plots.

Clustering analysis uses resemblance or dissemblance measures between the samples to be analyzed. The goal of this analysis is to obtain a symbolic description of the data and an identification pattern [44]. The most commonly used clustering algorithm is the hierarchical

http://dx.doi.org/10.5772/intechopen.77160

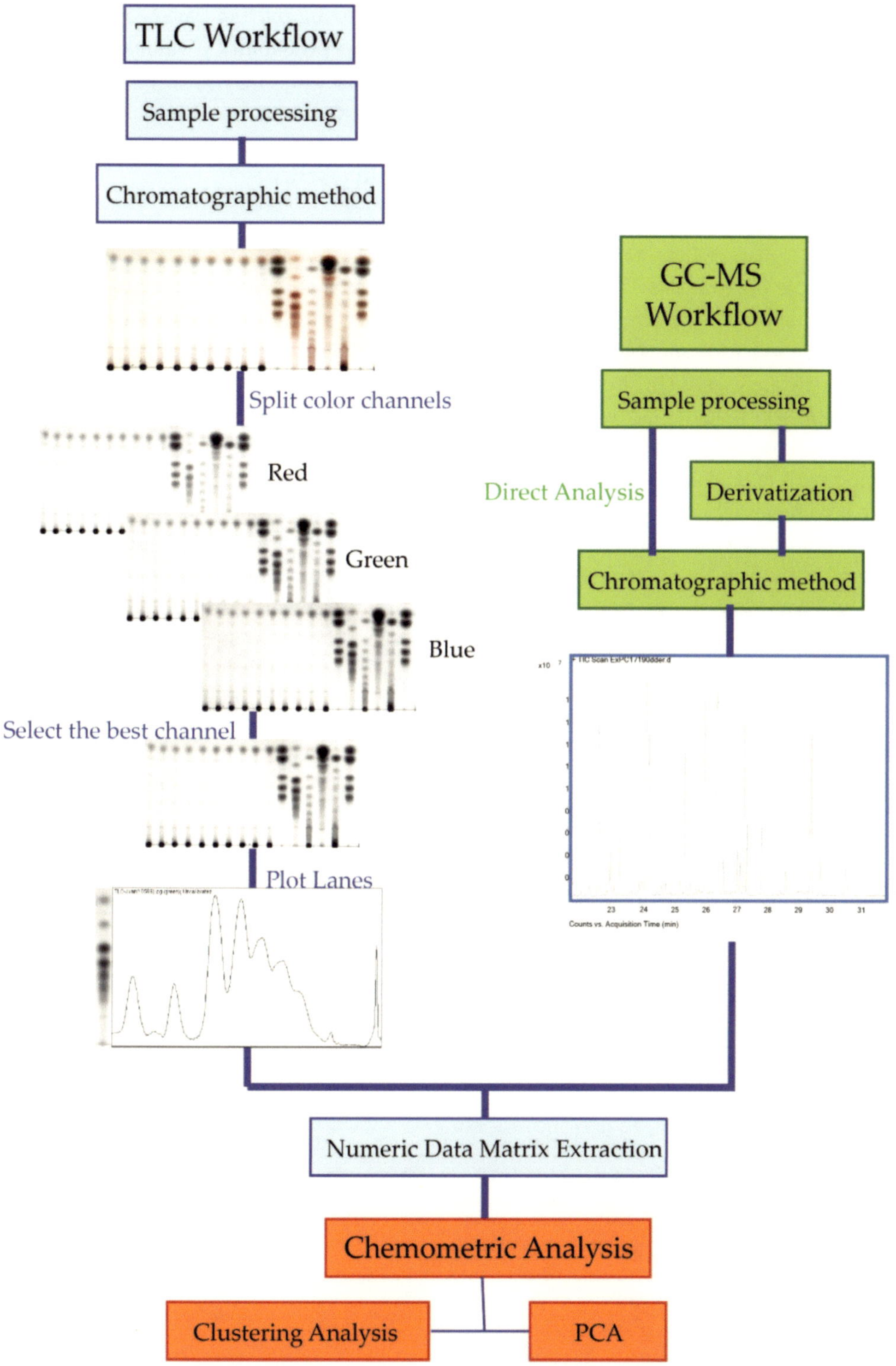

Figure 2. Workflow for chemometric analysis of TLC and GC-MS data.

method [44, 45]. For a hierarchical clustering procedure, the pvclust package in R could be used to provide approximately unbiased and bootstrap probability p values [46]. Other R packages could be also used to improve the appearance and analysis of the dendrograms [47]. In addition, model validation can be performed by an internal or external method [48].

4.3. Chemometrics-based TLC and GC-MS analysis in practice

4.3.1. Example of chemometrics-based TLC analysis

This example illustrates a TLC method using image analysis densitometry but with some consideration is also applicable to slit-scanning densitometry too. Four extracts of *Agave* (S2, S3, and S4) and *Cichorium* (S1) species containing considerable amounts of carbohydrates (mainly fructans) were chromatographed and derivatized by using diphenylamine-aniline-phosphoric acid reagent. Using this derivatization step, fructose and fructans appear as reddish spots, whereas glucose and maltooligosaccharides as bluish ones. After chromatographic development and visualization, the acquisition of an image of the TLC plate was done by using a commercial image densitometer TLC-Visualizer 2 (CAMAG), as shown in **Figure 3**. Once acquired, the TLC image was processed in the free software ImageJ (National Institutes of Health, NIH), and it was split into red, green, and blue channels. For this example, the red and blue channels were selected because the spots of interest were enriched and appear more defined in these channels. For noise removal, a median filter with an appropriate resolution

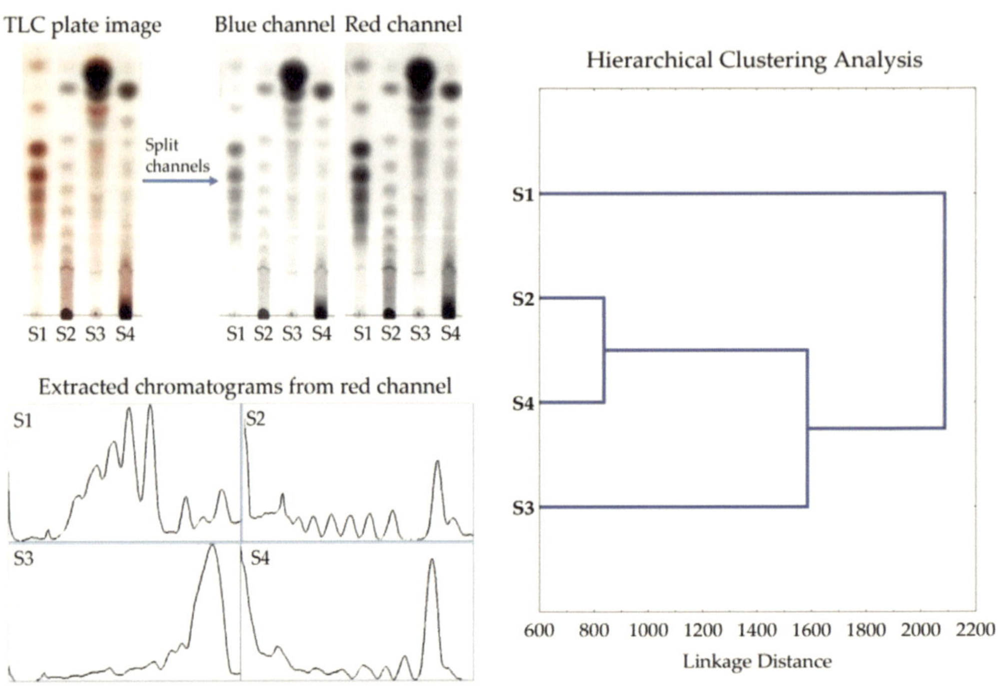

Figure 3. Hierarchical clustering analysis of oligosaccharide composition of four plant extracts: a TLC plate image with four samples (S1–S4) derivatized with diphenylamine-aniline-phosphoric acid reagent, the split color channel into blue and red, and the extracted chromatograms obtained for the red channel lanes.

was applied. A track containing the component(s) of interest was selected, designated as lane, and then plotted. The numeric data from each plot could be extracted to select exclusively the area of the plot line (avoid selecting any text or another data). In this study, a Cartesian graph was also obtained to extract single numeric data matrix for further chemometric analysis [49–51]. Besides the use of retention factor/relative abundance matrix, the relative peak area of all or part of the mixture components could also be used to construct data matrices.

The data matrix was used to construct a hierarchical clustering analysis with the software STATISTICA (StatSoft, Inc.), as displayed in **Figure 3**. This analysis allows to group the samples according to their monosaccharide and oligosaccharide composition without identification requirement. It is observed that the most closely related samples are S2 and S4, which form a clade and have the smallest distance among all the samples. S3 appears as sister of the S2–S4 clade, but with a distance equivalent to four times the distance between S2 and S4. S1 appears as an external group of the clade formed by S2–S4–S3, having a greater distance as compared to the other samples. According to this dendrogram, it is possible to conclude that based on the carbohydrate composition, S2 and S4 have roughly similar compositions among all the samples, S1 has the most different composition, and S3 has a composition most similar to S2–S4 than S1; this corresponds with the vegetal origin of the samples. It is observable that different polymerization degrees of the carbohydrates correspond with *Rf* values, that is, the higher the *Rf* value, the lower the polymerization degree. Thus, besides the grouping according to the type of carbohydrate (spots' color), the classification is also influenced by the size of carbohydrate molecules. The advantage of using chemometrics to analyze these TLC data consists of avoiding wasting time in compound identification and standard-based quantification, making the grouping and sample classification a fast but robust process.

4.3.2. *Example of chemometrics-based GC-MS analysis*

To demonstrate the application of the GC-MS methods herein described, a dataset obtained from the monosaccharide analysis of plant tissue extracts by GC-EIMS was studied. A total of nine samples obtained from three tissues by triplicate measurement (A = tissue 1, B = tissue 2, and C = tissue 3) were analyzed, and the total ion chromatograms were processed to extract the 204 and 217 m/z ion chromatograms with the aim of filtering data for silylated monosaccharides. After that, the chromatographic data were extracted and the numeric data matrix was subjected to principal component analysis. The resulting PCA data are shown in **Figure 4**; principal component 1 (PC1) and principal component 2 (PC2) explain 61.6 and 16.3% of the total variance, respectively. It can be observed that all samples are grouped according to the tissue from which they come. Samples of tissue B are closer, and samples of tissue C are more scattered. For B and C, the PC1 has a major effect on the data dispersion, whereas for A, the PC2 does the same thing. This nondirected analysis, in which no compounds have been identified, allowed grouping the data according to their origin, demonstrating the strength of the chemometric analysis. Although, in this example, only nine samples are used and their origin is known, this methodology has sufficient sturdiness to be applied to datasets of tens or even hundreds of samples with diverse sources as long as the same methodology is applied to all samples. In this chemometric analysis, our samples were grouped according to their monosaccharide content in a simple way, as well as for the above TLC chemometric analysis, avoiding wasting time in compound identification and standard-based quantification.

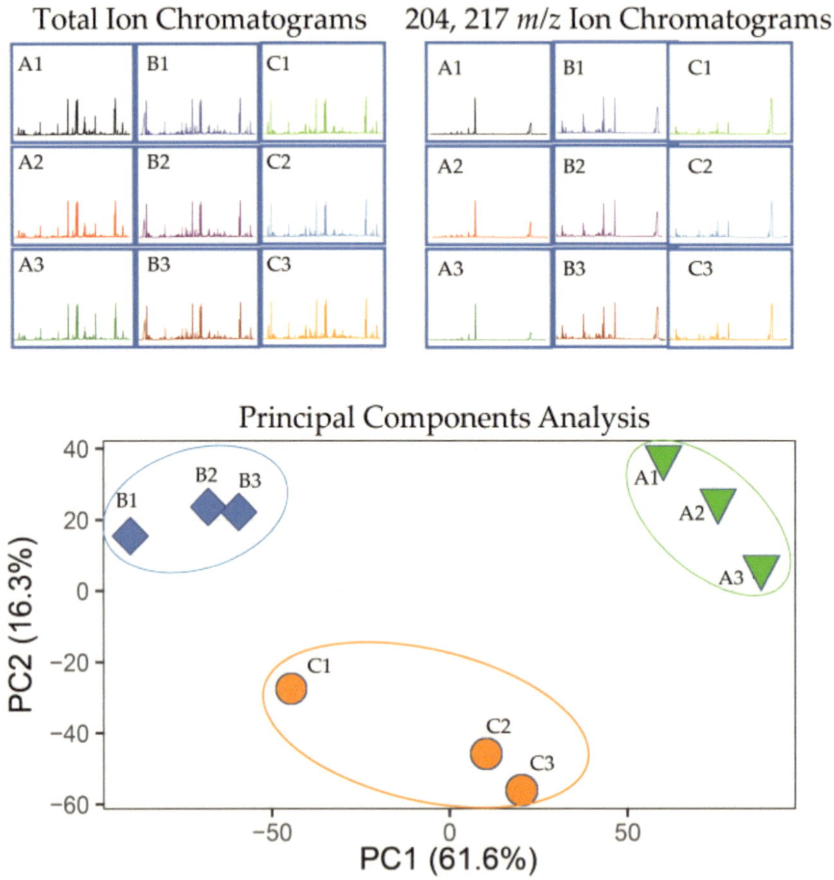

Figure 4. Chemometric analysis of GC-EIMS data of monosaccharides in plant tissues: the GC-EIMS total ion chromatograms (TICs), chromatograms obtained from the TIC extraction of the 204 and 217 *m/z* ions, and PCA constructed with the numeric data obtained from extracted chromatograms. Letter labels correspond to the tissue type and a label numeric fraction represents the sample replicate.

5. Conclusions

This chapter represents a practical guide for chemometric analysis of amino acids, short-chain fatty acids, and monosaccharides by thin-layer chromatography and gas chromatography coupled to mass spectrometry in complex mixtures. Furthermore, it provides a workflow to convert chromatographic data to numerical values, for which are described some chemometric analysis methods such as principal component analysis and clustering analysis.

Conflict of interest

Authors declare no conflict of interest.

http://dx.doi.org/10.5772/intechopen.77160

Author details

Juan Vázquez-Martínez and Mercedes G. López*

*Address all correspondence to: mercedes.lopez@cinvestav.mx

Center for Research and Advanced Studies of IPN Campus Irapuato (Centro de Investigación y de Estudios Avanzados del IPN Unidad Irapuato), Guanajuato, Mexico

References

[1] Deming SN. Chemometrics: An overview. Clinical Chemistry. 1986;**32**:1702-1706

[2] Roussel S, Preys S, Chauchard F, Lallemand J. Multivariate data analysis (chemometrics). In: O'Donnell CP, Fagan C, Cullen PJ, editors. Process Analytical Technology for the Food Industry. 1st ed. New York, USA: Springer-Verlag; 2014. pp. 7-17

[3] Gallo M, Ferranti P. The evolution of analytical chemistry methods in foodomics. Journal of Chromatography A. 2016;**1428**:3-15

[4] Siddiqui MR, AlOthman ZA, Rahman N. Analytical techniques in pharmaceutical analysis: A review. Arabian Journal of Chemistry. 2017;**10**:S1409-S1421

[5] Guidetti R, Beghi R, Giovenzana V. Chemometrics in food technology. In: Varmuza K, editor. Chemometrics in Practical Applications. 1st ed. Rijeka, Croatia: InTech; 2012. pp. 2017-2052

[6] Ciura K, Dziomba S, Nowakowska J, Markuszewski MJ. Thin layer chromatography in drug discovery process. Journal of Chromatography A. 2017;**1520**:9-22

[7] Sherma J, Fried B, editors. Handbook of Thin Layer Chromatography. 1st ed. New York, USA: Marcel Dekker, Inc.; 2003. 997 p

[8] Maloney MD. Carbohydrates. In: Sherma J, Fried B, editors. Handbook of Thin Layer Chromatography. 1st ed. New York, USA: Marcel Dekker, Inc.; 2003. pp. 579-610

[9] Sherma J. Biennial review of planar chromatography: 2013-2015. Journal of AOAC International. 2016;**99**:323-331

[10] Stroka J, Spangenberg B, Anklam E. New approaches in TLC-densitometry. Journal of Liquid Chromatography and Related Technologies. 2002;**25**:1497-1513

[11] Schönfled P, Wojtczak L. Short- and medium-chain fatty acids in energy metabolism: The cellular perspective. Journal of Lipid Research. 2016;**57**:943-954

[12] Nishi Y, Hiejima H, Hosoda H, et al. Ingested medium-chain fatty acids are directly utilized for the acyl modification of ghrelin. Endocrinology. 2005;**146**:2255-2264

[13] Koh A, De Vadder F, Kovatcheva-Datchary P, Bäckhed F. From dietary fiber to host physiology: Short-chain fatty acids as key bacterial metabolites. Cell. 2016;**165**:1332-1345

[14] Pyka A, Bober K. Visualizing agents for short-chain fatty acids in TLC. Journal of Planar Chromatography—Modern TLC. 2005;**18**:145-150

[15] Xie J, Schultz PG. Adding amino acids to the genetic repertoire. Current Opinion in Chemical Biology. 2005;**9**:548-554

[16] Wagner I, Musso H. New naturally occurring amino acids. Angewandte Chemie International Edition. 1983;**22**:816-828

[17] Mohammad A, Moheman A, El-Desoky GE. Amino acid and vitamin determinations by TLC/HPTLC: Review of the current state. Central European Journal of Chemistry. 2012;**10**:731-750

[18] Bhawani SA, Mohamad-Ibrahim MN, Salaiman O, et al. Thin-layer chromatography of amino acids: A review. Journal of Liquid Chromatography and Related Technologies. 2012;**35**:1497-1516

[19] Sleckman BP, Sherma J. A comparison of amino acid separations on silica gel, cellulose, and ion exchange thin layers. Journal of Liquid Chromatography. 1982;**5**:1051-1068

[20] Dolowy M, Pyka A. Application of TLC, HPLC and GC methods to the study of amino acid and peptide enantiomers: A review. Biomedical Chromatography: BMC. 2014;**28**:84-101

[21] Dale T, Court W. Improved separation of amino acids by thin-layer chromatography. Chromatographia. 1981;**14**:617-620

[22] Danishefsky SJ, DeNinno MP. Totally synthetic routes to the higher monosaccharides. Angewandte Chemie International Edition English. 1987;**26**:15-23

[23] Lee KY, So JS, Heo TR. Thin layer chromatographic determination of organic acids for rapid identification of bifidobacteria at genus level. Journal of Microbiological Methods. 2001;**45**:1-6

[24] Sinhababu A, Basu S, Dey H. Modified ninhydrin reagents to detect amino acids on TLC plates. Research on Chemical Intermediates. 2015;**41**:2785-2792

[25] Cheong K, Wu D, Hu D, et al. Comparison and characterization of the glycome of *Panax* species by high-performance thin-layer chromatography. Journal of Planar Chromatography—Modern TLC. 2014;**27**:1-5

[26] Mancilla-Margalli NA, López MG. Water-soluble carbohydrates and fructan structure patterns from *Agave* and *Dasylirion* species. Journal of Agricultural and Food Chemistry. 2006;**54**:7832-7839

[27] Ferey J, Da Silva D, Bravo-Veyrat S, et al. Validation of a thin-layer chromatography/densitometry method for the characterization of invertase activity. Journal of Chromatography A. 2016;**1477**:108-113

[28] Waksmundzka-Hajnos M, Sherma J, Kowalska T, editors. Thin Layer Chromatography in Phytochemistry. 1st ed. Boca Raton, USA: CRC Press; 2008. pp. 255-276

[29] Gutierrez-Villagomez JM, Vázquez-Martínez J, Ramírez-Chávez E, et al. Analysis of naphthenic acid mixtures as pentafluorobenzyl derivatives by gas chromatography-electron impact mass spectrometry. Talanta. 2017;**162**:440-452

[30] Hess AVI. Digitally enhanced thin-layer chromatography: An inexpensive, new technique for qualitative and quantitative analysis. Journal of Chemical Education. 2007;**84**:842-847

[31] Settle F, Pleva M, editors. Gas Chromatography-Mass Spectrometry, a Knowledge Base. 1st ed. Vol. 12. Amsterdam: Elsevier Science Publishers B.V.; 1989. 287 p

[32] García-Villalba R, Giménez-Bastida JA, García-Conesa MT, Tomás-Barberán FA, Espín JC, Larrosa M. Alternative method for gas chromatography-mass spectrometry analysis of short-chain fatty acids in faecal samples. Journal of Separation Science. 2012;**35**:1906-1913

[33] Orata F. Derivatization reactions and reagents for gas chromatography analysis. In: Ali Mohd M, editor. Advanced Gas Chromatography—Progress in Agricultural, Biomedical and Industrial Applications. 1st ed. Rijeka, Croatia: InTech; 2012. pp. 83-108

[34] Ruiz-Matute AI, Hernández-Hernández O, Rodríguez-Sánchez S, Sanz ML, Martínez-Castro I. Derivatization of carbohydrates for GC and GC-MS analyses. Journal of Chromatography B. 2011;**879**:1226-1240

[35] Grimm F, Fets L, Anastasiou D. Gas chromatography coupled to mass spectrometry (GC–MS) to study metabolism in cultured cells. In: Koumenis C, Coussens L, Giaccia A, Hammond E, editors. Tumor Microenvironment. Advances in Experimental Medicine and Biology. Vol. 899. Cham: Springer; 2016. pp. 59-88

[36] Zweckmair T et al. Improved quantification of monosaccharides in complex lignocellulosic biomass matrices: A gas chromatography-mass spectrometry based approach. Carbohydrate Research. 2017;**446**:7-12

[37] Brereton RG. Applied Chemometrics for Scientists. Chichester, UK: John Wiley & Sons, Ltd; 2007. 379 p

[38] Yip LY, Yong Chan EC. Gas chromatography/mass spectrometry-based metabonomics. In: Issaq HJ, Veenstra TD, editors. Proteomic and Metabolomic Approaches to Biomarker Discovery. 1st ed. Waltham, USA: Academic Press; 2013. pp. 131-144

[39] Wallace IM, Higgins DG. Supervised multivariate analysis of sequence groups to identify specificity determining residues. BMC Bioinformatics. 2007;**8**:1-12

[40] Pereira AC, Reis MS, Saraiva PM, Marques JC. Analysis and assessment of Madeira wine ageing over an extended time period through GC-MS and chemometric analysis. Analytica Chimica Acta. 2010;**660**:8-21

[41] Lê S, Josse J, Husson F. FactoMineR: An R package for multivariate analysis. Journal of Statistical Software. 2008;**25**:1-18

[42] R-Core-Team. R: A Language and Environment for Statistical Computing [Internet]. 2015. Available from: https://www.r-project.org/ [Accessed: March 15, 2018]

[43] RStudio-Team. RStudio: Integrated Developmet for R [Internet]. 2016. Available from: https://www.rstudio.com/ [Accessed: March 15, 2018]

[44] Diday E, Simon C. Clustering analysis. In: Fu KS, editor. Digital Pattern Recognition. 2nd ed. Berlin, Germany: Springer-Verlag; 1980. pp. 47-92

[45] Messai H, Farman M, Sarraj-Laabidi A, Hammami-Semmar A, Semmar N. Chemometrics methods for specificity, authenticity and traceability analysis of olive oils: Principles, classifications and applications. Food. 2016;**5**:1-35

[46] Suzuki R, Shimodaira H. Pvclust: An R package for assessing the uncertainty in hierarchical clustering. Bioinformatics. 2006;**22**:1540-1542

[47] Galili T. Dendextend: An R package for visualizing, adjusting and comparing trees of hierarchical clustering. Bioinformatics. 2015;**31**:3718-3720

[48] Enot DP, Hass B, Weinberger KM. Bioinformatics for mass spectrometry-based metabolomics. In: Mayer B, editor. Bioinformatics for Omics Data—Methods and Protocols. 1st ed. New York, USA: Humana Press; 2011. pp. 351-375

[49] Milojković Opsenica D, Ristivojević P, Trifković J, Vovk I, Lušić D, Tešić Ž. TLC fingerprinting and pattern recognition methods in the assessment of authenticity of poplar-type propolis. Journal of Chromatographic Science. 2016;**54**:1-7

[50] Jain R, Singh R, Sudhaker S, Kumar Barik A, Kumar S. Coupling microextraction with thin layer chromatography-image processing analysis: A new analytical platform for drug analysis. Toxicology and Forensic Medicine. 2017;**2**:17-25

[51] Olech M, Komsta Ł, Nowak R, Cieśla Ł, Waksmundzka-Hajnos M. Investigation of antiradical activity of plant material by thin-layer chromatography with image processing. Food Chemistry. 2012;**132**:549-553

Chapter 3

Chromatographic Separation, Total Determination and Chemical Speciation of Mercury in Environmental Water Samples Using 4-(2-Thiazolylazo) Resorcinol-Based Polyurethane Foam Sorbent-Packed Column

Amal H. Al-Bagawi, Waqas Ahmad, Hassan Alwael,
Zeinab M. Saigl, Gharam I. Mohammed,
Yousry M. Moustafa, Eman A. Al-Harbi and
Mohammad S. El-Shahawi

Additional information is available at the end of the chapter

http://dx.doi.org/10.5772/intechopen.81206

Abstract

A simple method has been developed for quantitative retention of traces of mercury(II) ions from aqueous media using polyurethane foams (PUFs) loaded with 4-(2-thiazolylazo) resorcinol (TAR). The kinetics and thermodynamics of the sorption of mercury(II) ions onto PUFs were studied. The sorption of mercury(II) ions onto PUF follows a first-order rate equation with $k = 0.176 \pm 0.010$ min^{-1}. The negative values of ΔH and ΔS may be interpreted as the exothermic chemisorption process and indicative of a faster chemisorption onto the active sites of the sorbent. The sorption data followed Langmuir, Freundlich and Dubinin-Radushkevich (D–R) isotherm models. The D-R parameters β, K_{DR} and E were 0.329 mol^2 kJ^{-2}, 0.001 µmol g^{-1} and 1.23 ± 0.07 kJ/mol for the TAR-loaded PUFs, respectively. An acceptable retention and recovery (99.6 ± 1.1%) of mercury(II) ions in water at ≤10 ppb by the TAR-treated PUFs packed columns were achieved. A retention mechanism, involving absorption related to "solvent extraction" and an "added component" for surface adsorption, was suggested for the retention of mercury(II) ions by the used solid phase extractor. The performance of TAR-immobilized PUFs packed column in terms of the number (N), the height equivalent to a theoretical plate (HETP), the breakthrough and critical capacities of mercury(II) ion uptake by the sorbent packed column were found to be 50.0 ± 1.0, 1.01 ± 0.02 mm, 8.75 and 13.75 mg/g, respectively, at 5 mL/min flow rate.

Keywords: retention of mercury(II) ions, polyurethane foams sorbent, 4-(2-thiazolylazo) resorcinol (TAR), separation mechanism, column chromatography

1. Introduction

Pollution by trace and ultra-trace concentration of heavy metal ions (e.g. mercury, lead, cadmium, arsenic, etc.) in our surroundings (air, soil and water) is an environmental concern due to their toxic effects and accumulation throughout the food chain, leading to serious ecological and health problems [1]. Mercury among the most toxic heavy metals commonly found in the global environment such as lithosphere, hydrosphere, atmosphere and biosphere [1, 2]. It is highly toxic to living cells and a bioaccumulative toxin that attacks the central nervous and endocrine systems [3]. Excessive prolonged exposure to mercury can cause brain damage and, in extreme cases, death [3]. The maximum concentration of mercury in drinking water recommended by the World Health Organization (WHO) is 1 ppb [4]. It can exist in three oxidation states (0, I, II) and all are able to combine with most inorganic and organic ligands to form various complexes e.g. HgX_4^{-2} (where X = Cl, Br and I) and methyl mercury [5, 6]. Mercury compounds are mutagenic and teratogenic in nature [7], and organomercury ones are more toxic than inorganic complex species of mercury(I) or mercury(II).

Techniques for mercury removal include traditional precipitation and coagulation, ion-exchange, solvent extraction, ultra filtration, and adsorption [8]. In recent years, a great deal of attention has been paid to the determination of trace and ultratrace concentrations of mercury by accurate, low-cost and reliable methods without any complicated sample preparation [9]. The most common techniques in analysis of natural water samples are stripping voltammetry [10], gas chromatography [11], spectrophotometry [12, 13], X-ray fluorescence spectrometry [14], and dual-wavelength β-correction spectrophotometric determination and speciation of mercury(II) ions in water using chromogenic reagent 4-(2-thiazolylazo) resorcinol [15].

Polyurethane foams (PUFs) and foams immobilized with supporting solvent extractants, chelating agents, liquid ion exchangers, anchored extracting groups and powdered ion exchangers sorbents have been reported for pre-concentration and separation of various inorganic and organic species from different media [16–22]. The most distinctive features of PUFs as solid sorbents are their membrane-like structure which differentiates them from others [20–22]. The foam membranes act as sorbents, i.e. the ions and/or molecules to be separated or pre-concentrated are retained on/in the membrane-like structure of the foams as shown in **Figure 1** [17–19]. The volume-to-surface ratio of quasi-spherical membrane geometry in comparison with cylindrical or planar one are the most advantageous characteristics of PUFs. Moreover, the PUFs membranes offer a wider range of chemical modifications than normal bulky (granular) solids [17, 20]. The potentialities of the PUF solid quasi-spherical membrane geometry sorbent have been reported [21, 22].

Open cell PUFs are a broad class of polymers having urethane moieties [22–25]. This class of polymers represents one of the most important three-dimensional (3D) products commercially available for fabricating super hydrophobic adsorbents for many organic and inorganic complex species [24–31]. They have excellent hydrophobicity and oleophilicity required for solid phase absorbents for various polar and non-polar species [25–32]. Commercial PUFs also exhibit poor hydrophobicity to some extent, as required for their practical use for clean-up and recovery of organic (e.g. oil spell) and inorganic species [33–39]. Hence, surface

Figure 1. Typical microscopic picture of PUFs.

modification of PUFs is desired to enhance the wettability and hydrophobicity to make it a super hydrophobic and oleophilic absorbent [25–34]. Furthermore, the emergence of nanomaterials with excellent properties (particle size and surface area) also makes them ideal for being coupled with PUFs as a model platform towards a target species [37–41].

Recently, much attention has been paid towards thiazole azo compounds as they are sensitive chromogenic reagents and good indicators in acid–base titrations in addition to their being important complexing agents for trace metal ions in aqueous media [10, 40–47]. In continuation to our previous data on the retention of trace metal ions from aqueous media [23, 25, 26], the overall objectives of this chapter are to study: (i) the sorption profile of mercury(II) ions from aqueous media using PUFs physically immobilized with the reagent 4-(2-thiazolylazo) resorcinol (abbreviated as TAR, **Figure 2**) as low-cost solid phase extractor; (ii) the kinetic, thermodynamic and sorption models and mechanism of mercury(II) ion uptake by the used PUF sorbents physically treated with TAR; (iii) chromatographic separation of mercury(II) ions from aqueous media by TAR-immobilized PUF packed column; (iv) a simple, low-cost and precise method based upon TAR-PUF sorbent packed column for separation and sequential determination of mercury(II) ions at ultra-trace level in water; and (v) assigning the most jprobable retention mechanism of mercury(II) ion retention by the used SPE from aqueous media.

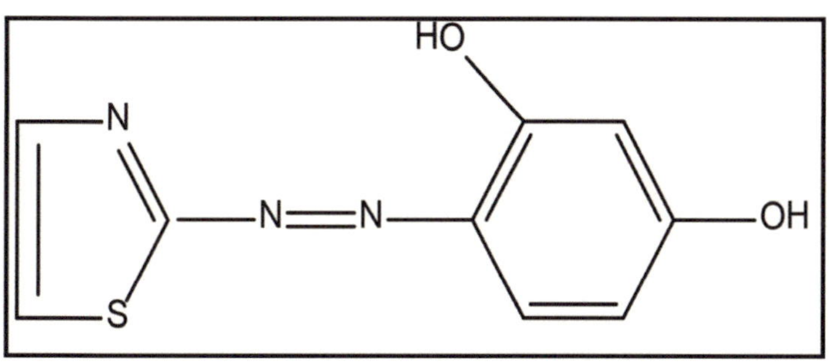

Figure 2. Chemical structure of 4-(2-thiazolylazo) resorcinol, TAR.

2. Experimental

2.1. Reagents and materials

Most of the chemicals were provided by Merck (Darmstadt, Germany). All chemicals were of analytical grade and used without further purification. All solutions were prepared in doubly de-ionized water and kept in a refrigerator. Plastic and glassware bottles were cleaned by soaking in dilute HNO_3 (10% w/v) and subsequently rinsed with distilled water prior to use. Working solutions (1–100 μg mL^{-1}) of mercury(II) ion were freshly prepared from the stock solution of mercury chloride ($HgCl_2$) (1000 μg mL^{-1}). Britton-Robinson buffers (pH 2–11) were prepared by mixing equimolar concentrations of acetic acid, phosphoric acid and boric acid (0.04 mol L^{-1}), and pH adjusting with NaOH (0.2 M) as reported [48].

2.2. Apparatus

A Perkin Elmer Elan DRC II ICP-MS system (USA) was used under the optimal default conditions. A Perkin Elmer ICP-OES system (Optima 4100 DC Shelton, CT, USA) was operated at the optimum operational parameters for mercury determination (**Table 1**). A Corporation Precision Scientific mechanical shaker (Chicago, CH, USA) with shaking 10–250 rpm and glass columns (15 cm × 10 mm i.d) (**Figure 3**) were used, respectively. pH measurements were done with a Thermo Fisher Scientific Orion model 720 pH meter (Milford, MA, USA), and de-ionized water obtained from a Milli-Q Waters Plus system (Milford, MA, USA). A digital micropipette (Volac) was used for solution preparation.

2.3. Preparation of TAR-immobilized PUFs

PUFs cubes (10–15 mm edge) were cut from commercially available white foam sheets, washed and dried as directed previously [25]. TAR reagent in ethanol (50 mL, 0.01% w/v) was shaken with PUFs cubes (4–5 g) with efficient stirring for 30 min. TAR-treated PUFs cubes

Parameter	Unit
Rf power (kW)	1300 W
Plasma gas (Ar) flow rate	15.0 L min^{-1}
Auxiliary gas (Ar) flow rate	0.2 L min^{-1}
Nebulizer gas (Ar) flow rate	0.8 L min^{-1}
Pump rate	1.5 mL min^{-1}
Observation height	15 mm
Integration time	10 s
Wavelength	327 nm

Table 1. ICP-OES operational parameters for mercury determination.

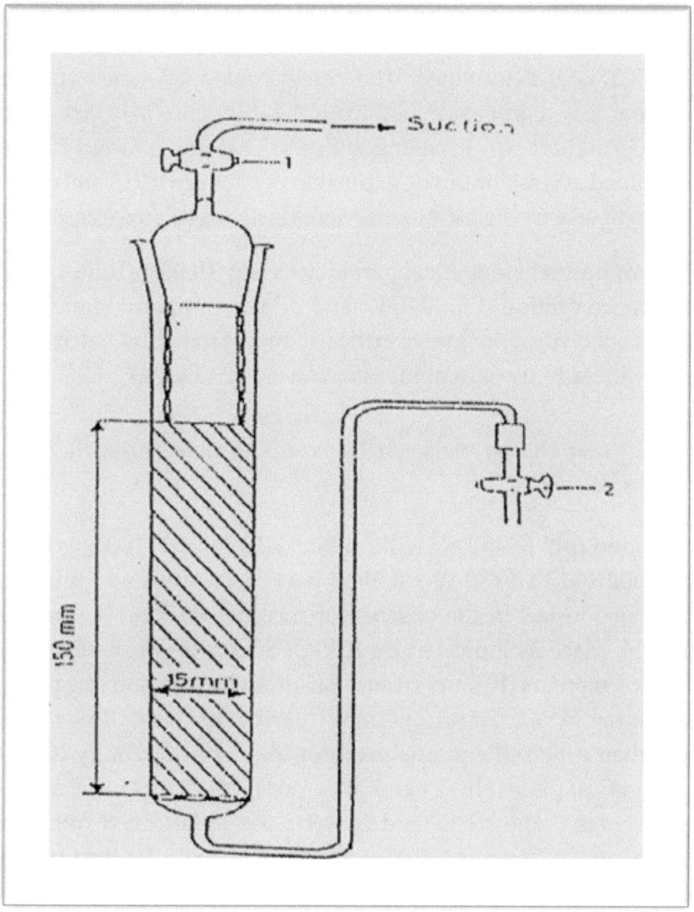

Figure 3. Preparation of PUFs packed column.

were squeezed and dried as reported earlier [25]. The amount of retained (q_e) reagent TAR onto the PUFs cubes was calculated by the following Equation [23, 25].

$$q_e = \frac{(C_b - C_a) \times V}{W} \tag{1}$$

where C_b and C_a are the initial and final concentrations of TAR in solution, respectively; V is the volume of the reagent solution (mL) and W is the mass (g) of the PUFs sorbent.

2.4. General procedures

2.4.1. Batch experiments

In a series of 250-mL conical flasks, an accurate weight (0.10 ± 0.01 g) of the TAR-immobilized PUFs was equilibrated with 200-mL Britton-Robinson buffers pH (2–12) containing mercury(II) ions 20 μg mL^{-1}. These solutions were shaken for 60 min at 25°C on a mechanical shaker. After phase separation, the aqueous phase was separated out by decantation and its mercury(II) ion concentration was ICP-OES determined. The concentration of mercury(II) ions retained on the foam solid sorbent was then determined from the difference between mercury(II) ion concentrations before (C_b) and after (C_a) shaking with the TAR-immobilized PUFs. The amount of mercury(II) ions retained at equilibrium q_e, extraction percentage (% E) and distribution ratio (D) of the mercury(II) ion uptake by the used foams were finally calculated as reported earlier [23–26].

Similarly, the effect of different analytical parameters [e.g. shaking time, cation size and monovalent metal ion concentration (Li^+, Na^+, K^+ and NH_4^+), sample volume, temperature, TAR and mercury(II) ion concentrations] were critically investigated by batch mode. The %E and D are the average of three independent measurements (RSD < 2%).

2.4.2. Chromatographic separation of mercury(II) ions on TAR-immobilized PUF sorbent-packed column

A 1-L aqueous solution (pH 6) spiked with mercury(II) ions (1–1000 μg L^{-1}) was percolated through TAR-immobilized PUFs (0.40 ± 0.01 g) packed columns at 5 mL min^{-1} flow rate. A blank sample was also tested in the absence of mercury(II) ions. A complete retention of mercury(II) ions took place as indicated by ICP-OES analysis of mercury in the effluent. A complete recovery of mercury(II) ions from TAR-immobilized sorbent packed column was achieved by percolating HNO_3 (20 mL, 1.0 mol L^{-1}) at 2 mL min^{-1} flow rate. Equal fractions of the eluate were then collected and analyzed for mercury content by ICP-OES. The height equivalent to a theoretical plate (HETP) and plate numbers (N) were calculated from the output of the chromatograms. The HETP and N were also determined from the breakthrough capacity curve (S-shaped) of mercury(II) ions at μg/mL level under the optimum condition of mercury(II) ion retention.

2.5. Analytical application

Tap and/or seawater samples (0.1–1.0 L) were collected and filtered through a 0.45-μm membrane filter; the solutions were pH adjusted to 6. All sample solutions were then spiked with different mercury(II) ion concentrations (0.001–1 μg mL^{-1}). They were percolated through TAR-loaded foam cubes packed columns at 2 mL min^{-1} flow rate. A complete sorption of mercury(II) ions was achieved as indicated by mercury analysis in the effluents. The amount of retained mercury(II) ions was recovered quantitatively with HNO_3 (20 mL, 1.0 mol L^{-1}) at 5 mL min^{-1} flow rate as noticed by ICP-MS analysis of the eluate.

3. Results and discussion

PUF is an excellent sorbent, which separates and pre-concentrates metal ions rather quickly compared with other solid sorbents [29–32]. It has been used in batch or dynamic techniques for the collection of a variety of chemical compounds, especially cations with harmful effects on humans, animals and plants e.g. Cd(II) and Hg(II).

3.1. Retention profile of mercury(II) ions onto the PUFs

The amount of mercury(II) ions extracted from the aqueous solution by the reagent TAR-immobilized PUFs was found to be pH dependent. The sorption profile of mercury(II) ions (20 μg mL^{-1}) from aqueous media by foams (0.10 ± 0.01 g) was investigated by employing Britton-Robinson buffers (pH 2–11). After shaking the test solutions with TAR-immobilized PUFs for 1.0 h at 25°C, the amount of mercury(II) ions remained in the aqueous media was measured by ICP-MS. The uptake percentage (%E) of mercury(II) ions from the aqueous media onto the TAR-immobilized PUFs sorbent was then calculated by the difference between mercury(II) ion concentrations before and after extraction.

As it can be seen in **Figure 4**, the sorption profile of mercury(II) ions by unloaded and TAR-loaded PUFs peaked at pH 6 and decreased progressively on further raising the solution pH. The observed decrease in mercury(II) ion uptake at pH > 6 is most likely attributed to the instability and/or hydrolysis of the complex species formed between mercury(II) complex species and the sorbent Hg-TAR. At pH < 6, the uptake of mercury(II) ions by the solid sorbents reduced. This reduction is probably attributed to TAR's low availability in ionized form and high availability in protonated amine group on the surface of TAR-functionalized PUFs [25].

A comparison between mercury(II) ion uptake from the aqueous media by TAR-treated PUF and untreated PUF (**Figure 4**) showed an excellent extraction performance for the former solid phase extractor. This behavior is most likely due to possible chelation between TAR on/in the surface and membrane of the PUF and mercury(II) ions in the aqueous media at pH 6 as expressed by the following equation:

$$Hg^{2+} aq + TAR_{PUF} \rightleftharpoons Hg\,(TAR)_{PUF} \qquad (2)$$

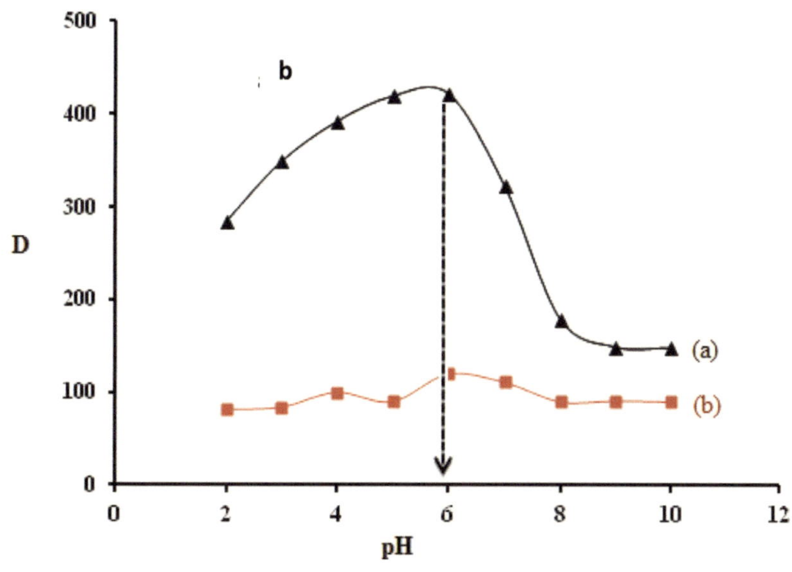

Figure 4. Plot of distribution ratio of mercury(II) ion uptake by TAR-treated PUFs (a) and unloaded PUFs (b) *versus* pH at 25°C after 1-h shaking.

The stability constants of the binding sites of the PUFs with mercury(II) ions were calculated using the Scatchard Eq. [25]:

$$\frac{n}{[Hg]} = K(n_i - n) \quad (3)$$

$$n = \frac{\text{weight of mercury bound to foam (g)}}{\text{weight of foam (g)}} \quad (4)$$

where K is the stability constant of mercury(II) ions on PUFs, n_i is the maximum concentration of mercury(II) ions sorbed by the available sites on the PUFs, and [Hg] is the equilibrium concentration of mercury(II) ions (mol L^{-1}). As shown in **Figure 5**, the curvature of the Scatchard plot demonstrates that more than a mercury(II) complex species could be formed and has its own unique formation constant [23, 25]. The stability constants log K_1 and log K_2 for the sorbed species derived from the respective slopes were 5.365 ± 0.200 and 4.216 ± 0.500. The calculated values of n_{i1} and n_{i2} were 0.022 ± 0.005 and 0.053 ± 0.010 mol g^{-1}, respectively. The values of the stability constants (log K_1 and log K_2) indicate that the sorption of this species took place readily on site K_1, most likely due to the ether group because this group has a greater stability than the amide group (site K_2) [25].

As demonstrated in **Figure 6**, the sorption of mercury(II) ions onto TAR-immobilized PUFs sorbents peaked at about 30 min and remained relatively constant after 60 min. The half-life ($t_{1/2}$) of the equilibrium sorption of mercury(II) ions onto TAR-immobilized PUFs sorbents from the aqueous phase was calculated from the plot of –log $C_b - C_a/C_b$ *versus* time (**Figure 7**). The $t_{1/2}$ value of mercury(II) ion uptake by TAR-PUF sorbent was found to be 2.00 ± 0.08 min. It means that most likely gel diffusion is not the only rate-controlling step for TAR-immobilized PUFs sorbent as reported for common ion exchange resins [23, 25]. Therefore, the kinetic

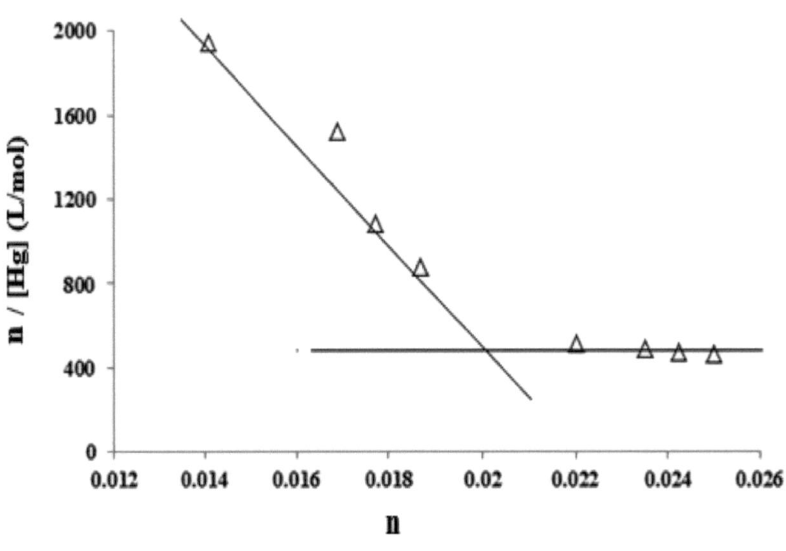

Figure 5. Scatchard plot of the sorption of mercury(II) ions from the aqueous media onto TAR-immobilized PUFs sorbents at 25°C after 1-h shaking.

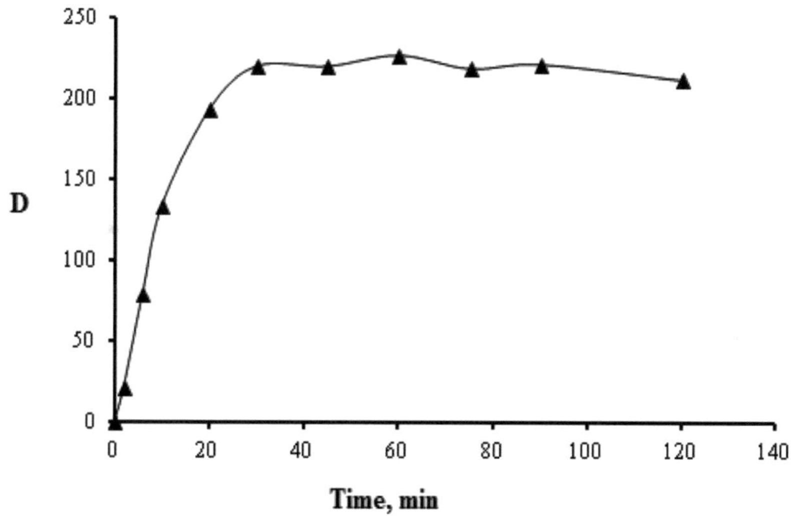

Figure 6. Effect of shaking time on the sorption of mercury(II) ions from the aqueous media onto TAR-immobilized PUFs sorbents at 25C after 1h shaking time at pH 6.

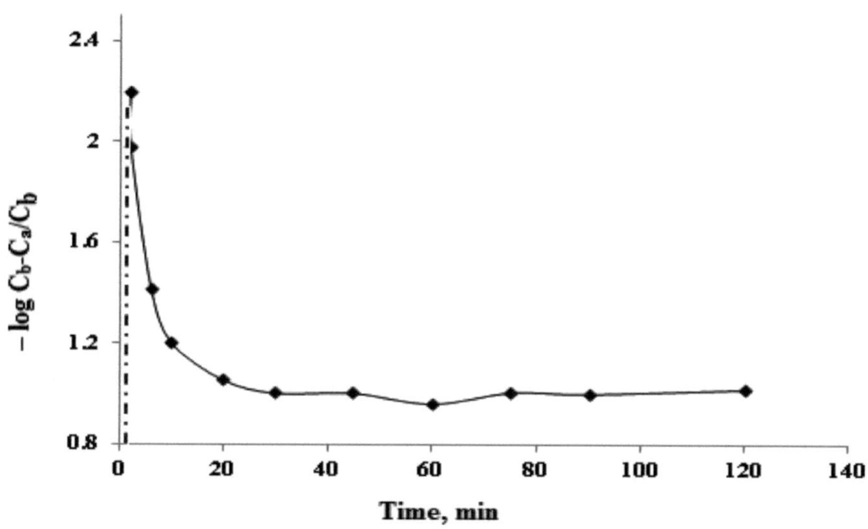

Figure 7. Effect of shaking time on the sorption of mercury(II) ions from the aqueous media onto TAR-loaded PUFs sorbents at 25°C after 1h shaking time at pH 6.

behavior of mercury(II) ion sorption onto the TAR-immobilized PUFs sorbent depends on the film diffusion and intraparticle diffusion that probably control the overall transport of mercury(II) ions in/on the sorbent membrane.

Figure 8 displays the effect of cation size of various monovalent alkali metal ions (Li^+, Na^+, K^+ and NH_4^+) as chloride salt (0–2% w/v) on mercury(II) ion uptake (20 ppm) from the aqueous media at the optimized pH 6 onto TAR-loaded PUFs sorbents (0.10 ± 0.01 g) after 1-h shaking. The sorption profile of mercury(II) ions increased with a decrease in the cation size and the overall retention step was in the following order:

$$Li^+ > Na^+ \sim NH_4^+ > K^+ \quad (5)$$

On the other hand, when monovalent cations were added to the extraction media (0.04–1.2% w/v), the reduction of a number of water molecules available to solvate mercury(II) ions forced the solvent phase out TAR-treated PUF sorbents [23, 25]. This finding is in good agreement with the data reported by El-Shahawi et al., clearly stating that some amount of free water molecules is preferentially used to solvate cations added leading to an analyte uptake enhancement [25]. The ion-dipole interaction of NH_4^+ with the oxygen sites of the PUFs was not a predominating factor in the extraction of mercury(II) ions. This effect is consistent with the "solvent extraction" mechanism with a salting-out agent.

Investigation on the effect of sample volume (20–500 mL) of mercury(II) solution (20 μg mL^{-1}) shows that the extraction percentage of mercury(II) ions from the aqueous media slightly decreased (10–15%) when using a sample volume > 200 mL. In contrast, no significant effect on the mercury(II) ion uptake was observed when varying TAR concentration (0.02–0.14% w/v) (**Figure 9**). As a result, TAR concentration was adopted at 0.05% (w/v) for subsequent work.

3.2. Kinetics of the mercury(II) ion sorption by TAR-treated PUFs

The retention of inorganic pollutants such as mercury(II) species from the aqueous media plays an important role in wastewater treatment. In the present study, data on mercury(II) ion uptake by the used solid sorbents were subjected to many kinetic models e.g. Morris-Weber [49], Lagergren pseudo-first order [50], Bhattacharya & Enkobachar [51], Reichenburg [52] and pseudo-second order rate models [53]. The Weber-Morris model can be mathematically expressed as follows [49]:

$$q_t = R_d (t)^{1/2} \quad (6)$$

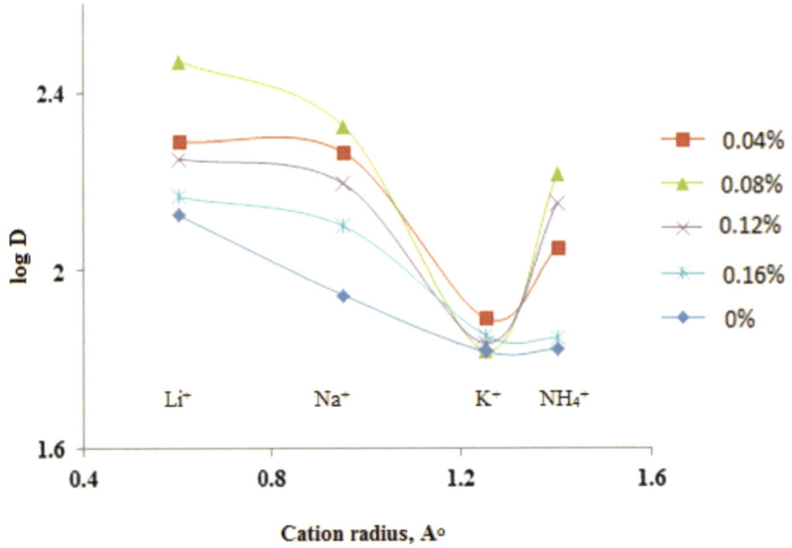

Figure 8. Effect of monovalent cation (Li^+, Na^+, K^+, NH_4^+) size and concentration (1% w/v) on the sorption profile of mercury(II) ions onto unloaded (a) and TAR-loaded PUFs (b) at 25°C after 1-h shaking.

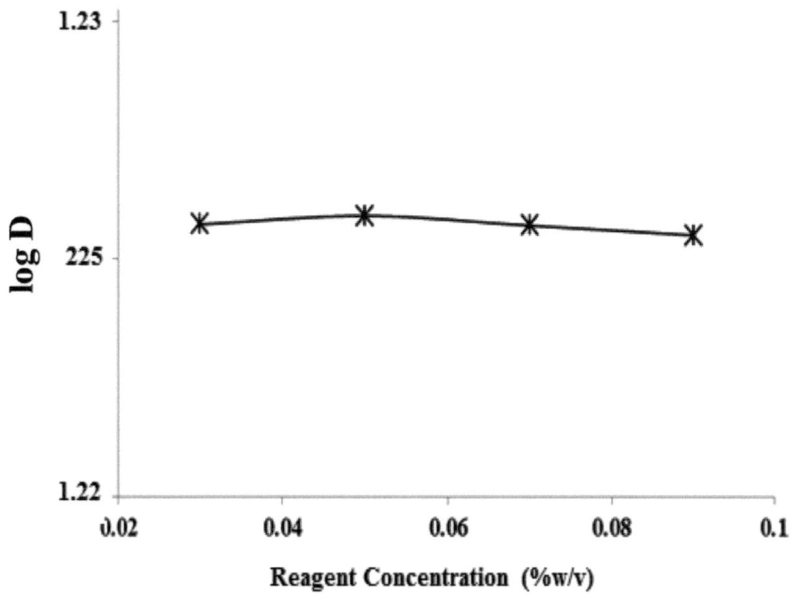

Figure 9. Effect of immobilized-TAR concentration (% w/v) of the loaded foams on mercury(II) ion uptake from the aqueous media pH 6 at 25°C after 1-h shaking.

where R_d is the rate constant of intraparticle transport in mg g^{-1} $min^{-1/2}$ and q_t is the sorbed mercury(II) ion concentration (mg g^{-1}) at time t (as shown in **Figure 10**). The plot of q_t *versus* time were found linear ($R^2 = 0.995$) in the initial stage for mercury(II) ion retention onto the sorbent up to 30.25 ± 0.03 min; on further increasing the shaking time, the straight line did not pass through the origin. It indicates that (i) intraparticle diffusion is not the only rate-controlling step for TAR-immobilized PUFs as in the case of common ion exchange resins [23, 25] and (ii) the retention step is mainly controlled by film diffusion at the early stage of extraction and in the second stage, the diffusion remains fairly constant when the pores volume of PUFs is exhausted [25].

The values of R_d computed from the two slopes in the initial and second stages of the Weber-Morris plots for TAR-loaded PUFs were found to be 0.707 ± 1.010 mg (g min)$^{-1}$ and 0.053 ± 0.020 mg (g min)$^{-1}$ with $R^2 = 0.995$ and 0.989, respectively. The change in the slope is most likely related to the existence of different pore sizes [44, 45], further confirming intraparticle diffusion as the rate-controlling step.

Moreover, the rate constant for the retention step of mercury(II) ion retention onto the tested solid sorbent was also evaluated in the light of Lagergren Equation [50]:

$$\log(q_e - q_t) = \log q_e - \frac{K_{Lager}}{2.303}\, t \qquad (7)$$

where q_e is the amount of mercury(II) ions sorbed at equilibrium per mass unit of sorbent (mmol g^{-1}), k_{Lager} is the first order overall rate constant for the retention process per min and t is the time in min. The value of K_{Lager} calculated from the linear plot of log ($q_e - q_t$) *versus* time (**Figure 11**) was found to be 0.176 ± 0.010 min^{-1}, suggesting first-order kinetics for the mercury(II) ion retention onto the sorbent. The value of K_{Lager} increases proportionally to the increasing

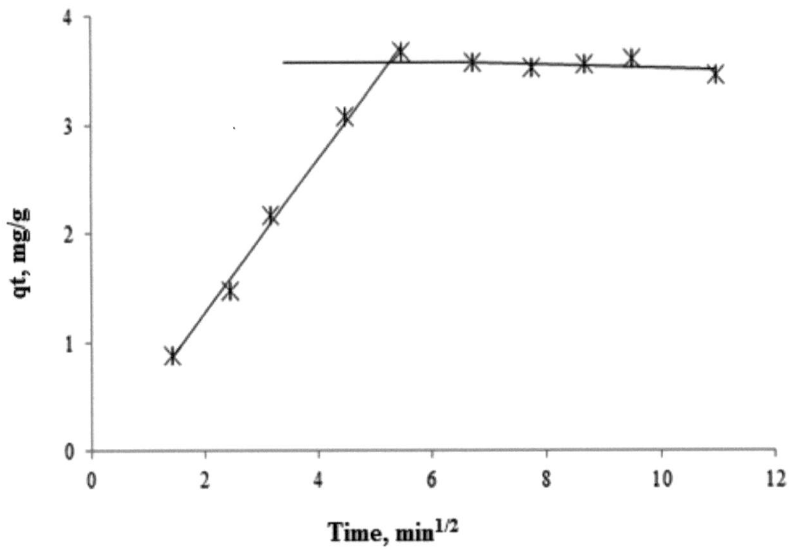

Figure 10. Weber–Morris plot of the sorbed mercury(II) ions onto TAR-immobilized PUFs sorbents versus square root of time at 25°C after 1-h shaking.

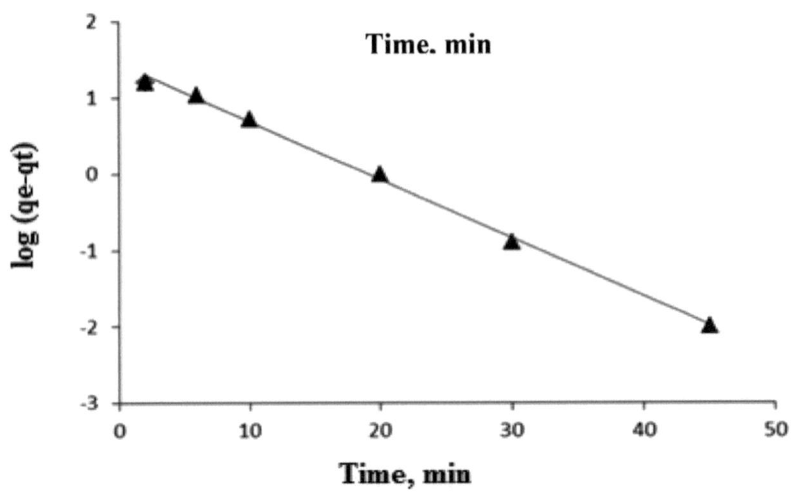

Figure 11. Lagergren plot of mercury(II) ions uptake onto TAR-immobilized PUFs sorbents from the aqueous media versus time at 25°C after 1-h shaking.

sorbate concentration, confirming a monolayer formation of mercury(II) ions onto the surface of the used sorbent as well as the first-order kinetic nature of the retention process [23, 25].

These results were also confirmed by Bhattacharya-Enkobachar kinetic model [51].

$$\log\left(1-U_{(t)}\right) = \frac{-K\,Bhatt}{2.303}\,t \qquad (8)$$

where $U_{(t)} = \frac{C_o - C_t}{C_o - C_e}$, K_{Bhatt} = overall rate constant (min^{-1}), t = time (min), C_t = concentration of mercury(II) ions at time t in μg mL^{-1}, C_e = concentration of mercury(II) ions at equilibrium in μg mL^{-1}. The value of K_{Bhatt} estimated from the linear region of the plot of log (1 – U_t) against time (**Figure 12**) was equal to 0.183 ± 0.002 min^{-1}. This value is quite close to that achieved by Lagergren model, and provides an additional indication of the first-order kinetics for mercury(II) ion retention onto the used sorbent.

The value of Bt, which is a mathematical function (F) of the ratio of the fractions sorbed at time t (q_t) and at equilibrium (q_e) in μmole g^{-1} (i.e. F = q_t / q_e), is calculated for each value of F by employing Reichenburg Equation [52, 53].

$$Bt = -0.4977 - 2.303 \log(1-F) \tag{9}$$

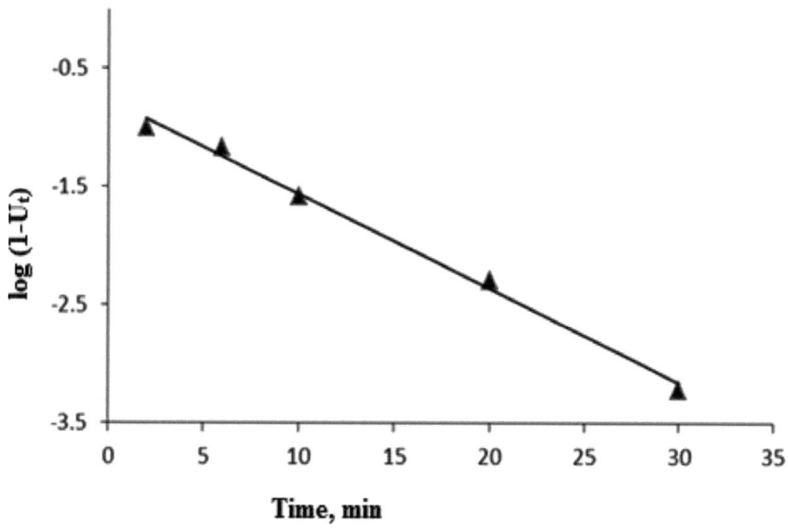

Figure 12. Bhattacharya-enkobachar plot of mercury(II) ion retention from the aqueous media onto TAR-loaded PUFs sorbents at 25°C after 1-h shaking.

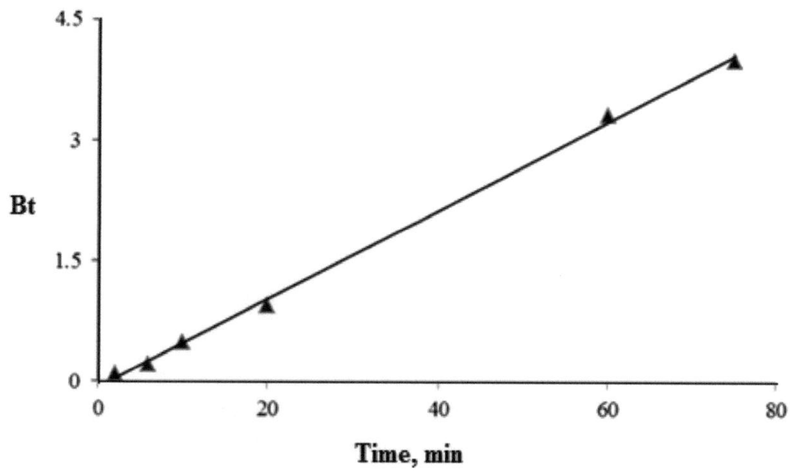

Figure 13. Reichenburg plot of the kinetics of mercury(II) ions retention onto TAR-loaded PUFs sorbents from the aqueous media at 25°C after 1-h shaking.

the plot of Bt *versus* time at 25°C for mercury(II) ion sorption was linear (R^2 = 0.998) up to 80 min (**Figure 13**). The straight line does not pass through the origin, indicating that particle diffusion mechanism is not responsible for mercury(II) ion sorption onto the sorbent. Thus, the uptake of mercury(II) ions onto the sorbent may consist of three steps: (i) bulk transport of mercury(II) ions in solution, (ii) film transfer involving diffusion of mercury(II) ions within the pore volume of TAR-treated PUFs sorbents and/or along the wall surface to the active sorption sites of the sorbent, and (iii) formation of Hg^{2+}-TAR chelate on/in the sorbent. This explains that fact that the sorption of mercury(II) ions on the interior surface is very rapid and hence it is not a rate-determining step.

3.3. Thermodynamic characteristics of the mercury(II) ion retention onto TAR-loaded PUFs sorbents

The sorption behavior of mercury(II) ions by unloaded and TAR-loaded PUFs sorbents was critically studied over a wide range of temperatures (293–353 K) to determine the characteristics of mercury(II) ion retention onto TAR-loaded PUFs sorbents at the established conditions. The thermodynamic parameters (ΔH, ΔS, and ΔG) were evaluated by using the following equations [23, 25]:

$$\text{In } K_c = \frac{-\Delta H}{RT} + \frac{\Delta S}{R} \quad (10)$$

$$\Delta G = \Delta H - T\Delta S \quad (11)$$

$$\Delta G = -RT \ln \text{Kc} \quad (12)$$

where ΔH, ΔS, ΔG, and T are the enthalpy, entropy, Gibbs free energy changes and temperature in Kelvin, respectively; and R is the gas constant (≈ 8.3 J K^{-1} mol^{-1}). K_C is the equilibrium constant depending on the fractional attainment (Fe) of the sorption process. The values of K_C for retention of mercury(II) ions from the aqueous media at equilibrium onto unloaded and TAR-loaded PUFs sorbents was calculated by employing the following equation:

$$K_C = \frac{Fe}{1 - Fe} \quad (13)$$

Figure 14 shows a reduction in the values of K_C with increasing temperature, revealing that the retention of mercury(II) ion by the sorbent is an exothermic process. The numerical values of ΔH, ΔS, and ΔG calculated from the slope and intercept of the linear plot were found to be – 41.5 ± 1.01 kJ mol^{-1}, – 132.8 ± 0.5 J mol^{-1} K^{-1} and – 1.62 kJ mol^{-1} (at 298 K), respectively. The ΔH, ΔS, and ΔG values found were equal to – 39.80 ± 0.01 kJ mol^{-1}, – 128.6 ± 0.3 J mol^{-1} K^{-1} and – 1.47 kJ mol^{-1} (at 298 K), respectively.

Considering the Van't Hoff equation in terms of the distribution ratio of mercury(II) ions, D, the following expression is obtained:

$$\log D = \frac{-\Delta \text{H}}{2.30\ RT} + 1/\text{n}\ C \quad (14)$$

http://dx.doi.org/10.5772/intechopen.81206

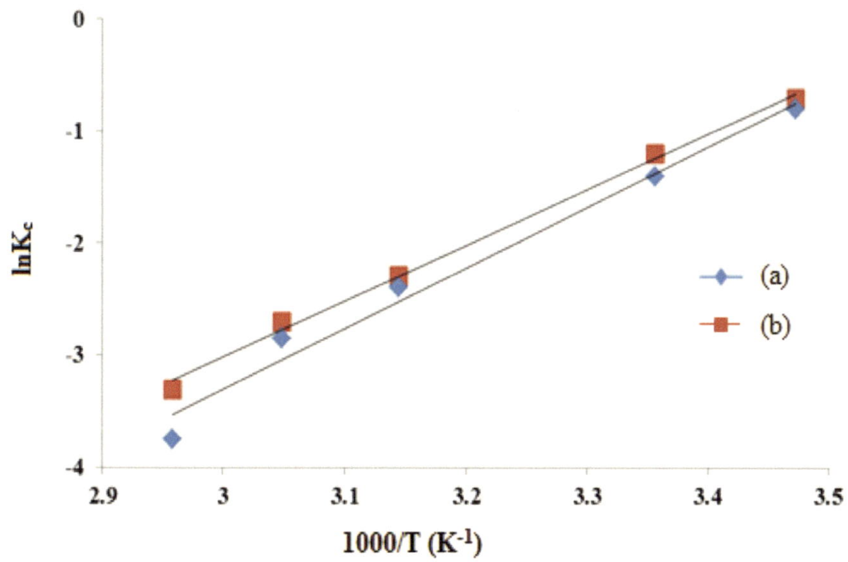

Figure 14. Plot of lnK_c of mercury(II) ion sorption versus 1000/T (K^{-1}) from the aqueous media pH 6 onto unloaded (a) and TAR-loaded PUFs sorbents (b) at 25°C after 1-h shaking.

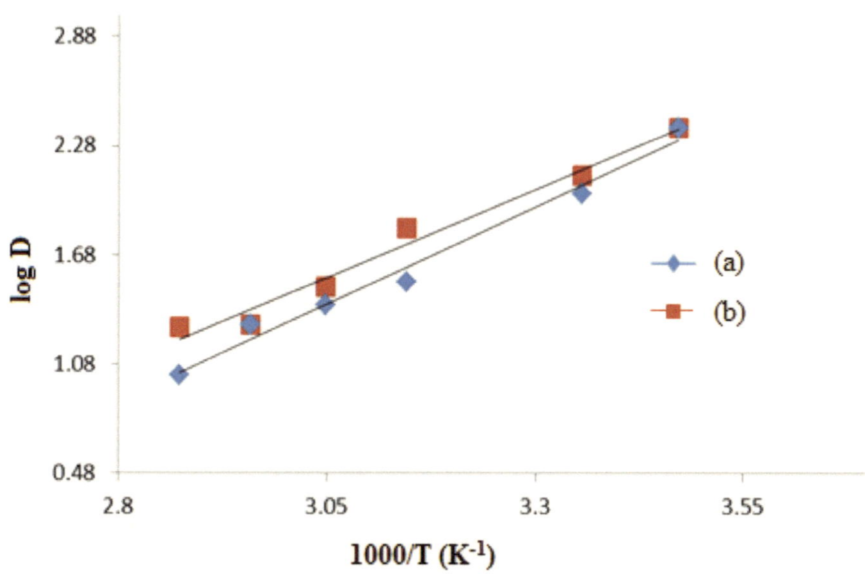

Figure 15. Van't-hoff plot for mercury(II) ion retention onto TAR-loaded PUFs (a) and unloaded sorbents (b) at 25°C after 1-h shaking.

where C is a constant. The values of D of mercury(II) ion retention from the aqueous media pH 6 onto unloaded and TAR-loaded PUFs sorbents decreased on raising temperature. The plots of log D *versus* 1000/T were linear (**Figure 15**). The calculated values of ΔH for mercury(II) sorption were found to be – 38.4 and – 40.1 kJ mol^{-1} onto unloaded and TAR-loaded PUFs sorbents, respectively.

The negative value of ΔH and the values of D and K_C reflect the exothermic nature of mercury(II) ion uptake by the employed solid PUFs as well as the non-electrostatic bonding formation between the sorbent and adsorbate. The negative values of ΔS may be indicative of the moderate sorption step of mercury(II) complex ion associate and charge ordering without a compensatory disordering of the sorbed ion associate onto the used sorbents. The motion of mercury(II) ions is more restricted in PUF membrane than in solution. Since the sorption process involves a decrease in free energy, the ΔH value is expected to be negative as confirmed above.

The negative value of ΔG at 295 K indicates the spontaneous and physical sorption characteristics of mercury(II) ion retention onto PUFs. An increase in the values of ΔG with temperature may be due to the spontaneous nature of the sorption step, and mercury(II) ion uptake is more favorable at low temperature confirming the exothermic sorption process. The energy of urethane nitrogen and/or ether – oxygen sites of the PUFs with increasing temperature minimizes a possible interaction between the active sites of PUFs and mercury(II) complex ion associates, resulting in a lower sorption percentage of the analyte. Therefore, a sorption mechanism involving "solvent extraction" is most likely responsible for the retention step.

3.4. Sorption isotherms of the mercury(II) ion retention by TAR-treated PUFs sorbents

The sorption characteristics of mercury(II) ions over a wide range of equilibrium concentrations (1–80 μg mL^{-1}) from the aqueous solution pH 6 onto unloaded and TAR-loaded PUFs were studied. As shown in **Figure 16**, at low or moderate analyte concentrations the amount of mercury(II) ions retained on the sorbent varied linearly with the amount of mercury(II) ions remained in the bulk aqueous solution. The equilibrium was only established for mercury(II) ion rich aqueous phase, confirming the first-order kinetics of the sorption step. A relatively reasonable sorption capacity of mercury(II) ions onto unloaded and TAR-loaded PUFs sorbents (as predicted from the sorption isotherm) was found to be greater than 16.70 ± 0.02 and 24.40 ± 0.07 mg g^{-1}, respectively. The K_d values decreased on raising mercury(II) ion concentration when the sorbent membrane became more saturated with mercury(II) ions within 15–20 min of shaking (**Figure 17**).

The retention behavior of mercury(II) ions from the aqueous media onto the used sorbents was subjected to Freundlich, Langmuir and Dubinin-Radushkevich (D-R) isotherm models over a wide range of equilibrium concentrations by means of the linear least squares fitting technique. The Freundlich model [54] is expressed in the following form:

$$\log C_{ads} = \log A + \frac{1}{n} \log C_e \quad (15)$$

where A and $\frac{1}{n}$ are Freundlich parameters related to the maximum sorption capacity of solute (mol g^{-1}) and C_{ads} is the sorbed mercury(II) ion concentration per mass unit of TAR-immobilized PUFs sorbents (mol g^{-1}) at equilibrium. Plot of $\log C_{ads}$ *vs.* $\log C_e$ (**Figure 18**) for the mercury(II) ion retention onto TAR-immobilized PUFs was linear (R^2 = 0.993) over the entire concentration range of the analyte, indicating a good fit for the experimental data. The Freundlich sorption isotherm encompasses the heterogeneity of the surface and exponential distribution of the sites and their energies. The Freundlich parameters A and 1/n computed from

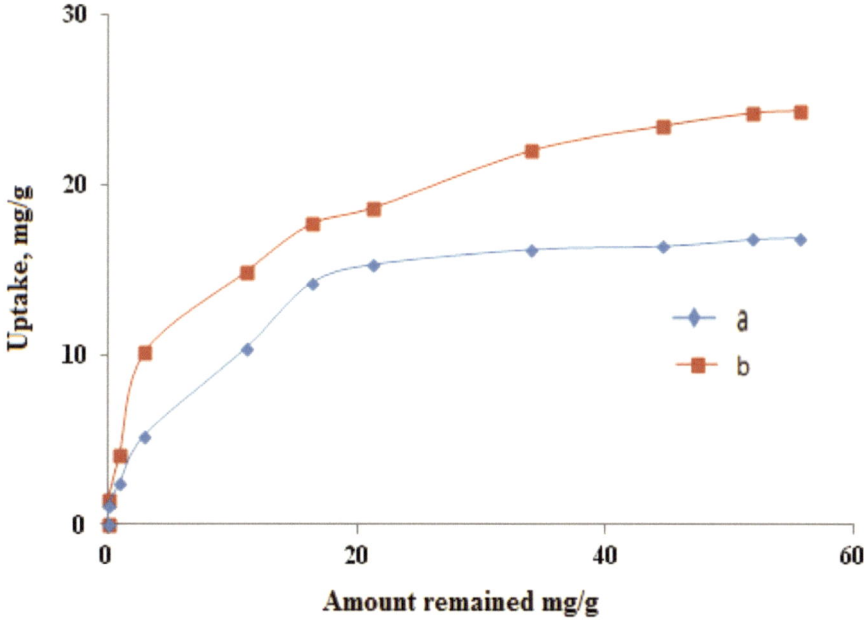

Figure 16. Sorption isotherms of the mercury(II) ion uptake from the aqueous media at optimal conditions onto unloaded (a) and TAR-loaded PUFs sorbents (b) at 25°C after 1-h shaking.

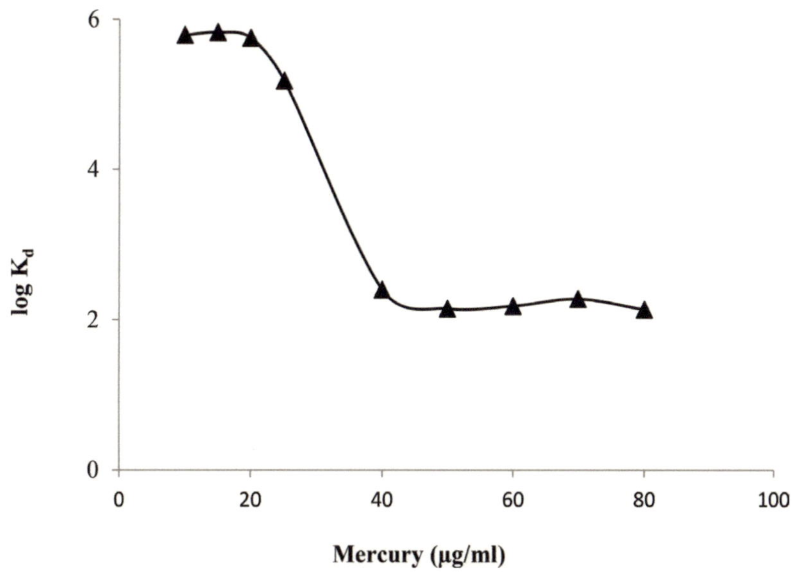

Figure 17. Plot of log K_d of the mercury(II) ion sorption *versus* mercury(II) ion concentration in the aqueous media pH 6 onto TAR-loaded PUFs sorbents.

the intercept and slope of the linear plot (**Figure 18**) were found to be 0.251 ± 0.100 $mmol.g^{-1}$ and 0.25 for TAR-loaded PUFs and 0.501 ± 0.300 $mmol.g^{-1}$ and 0.355 for unloaded PUFs. The value of $1/n < 1$ revealing favorable sorption of mercury(II) ions onto the tested solid sorbents.

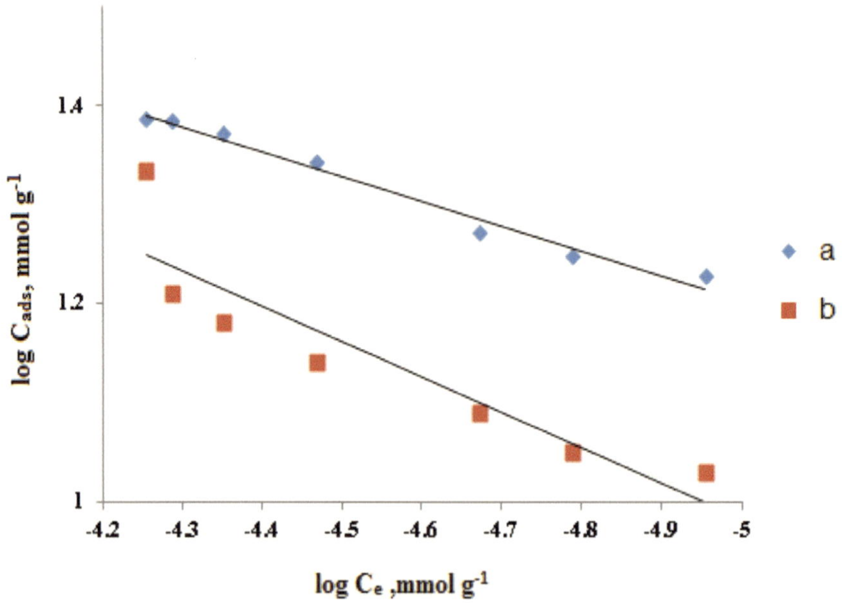

Figure 18. Freundlich sorption isotherms of the mercury(II) ion uptake from the aqueous solution at optimal conditions onto unloaded (a) and TAR-loaded PUFs sorbents (b) at 25°C after 1-h shaking.

At lower equilibrium concentration, the sorption capacity slightly decreased and the isotherm does not predict any saturation of the surface of the PUFs the adsorbate. Thus, an infinite surface coverage is mathematically predicted and a physisorption on the surface of the sorbents is expected.

Langmuir sorption isotherm model is expressed in the following linear form [55].

$$\frac{C_e}{C_{ads}} = \frac{1}{Q_b} + \frac{Ce}{Q} \tag{16}$$

where C_e is the equilibrium concentration (mg L^{-1}) of mercury(II) ions. The constant Q is the Langmuir parameter related to the maximum adsorption capacity of solute per mass unit of adsorbent required for a monolayer coverage of the surface; and b is an equilibrium constant related to the binding energy of solute sorption that is independent of temperature. The plot of C_e/C_{ads} *versus* C_e over the entire mercury(II) ion concentration was linear **(Figure 19)**, confirming the validity of Langmuir adsorption model for analyte uptake onto the sorbent. The parameters Q and b calculated from the slop and intercept of Langmuir plot were found to be 4.89 ± 0.01 mmol g^{-1} and 2.61 L mol^{-1} for TAR-loaded PUFs, and 3.20 ± 0.04 mmol g^{-1} and 0.74 L mol^{-1} for unloaded PUFs.

The Dubinin-Radushkevich (D-R) isotherm model [56] is postulated within the adsorption space close to the adsorbent surface. The D-R isotherm can be linearized as follows:

$$\ln C_{ads} = \ln K_{DR} - \beta \varepsilon^2 \tag{17}$$

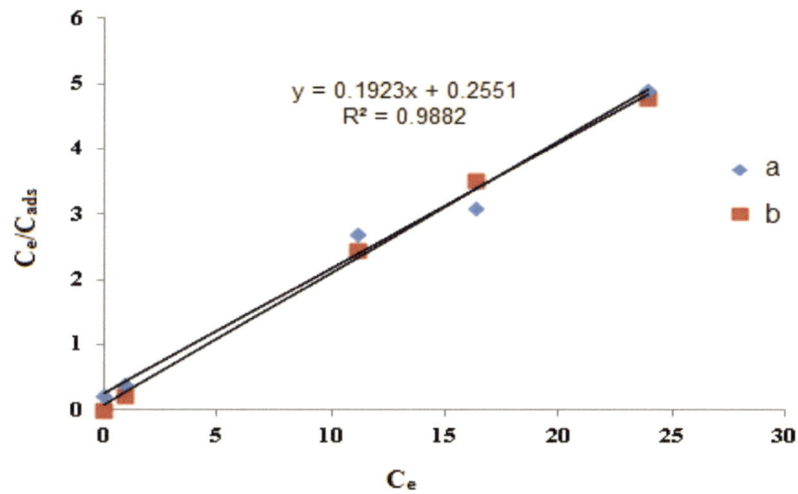

Figure 19. Langmuir sorption isotherm of the mercury(II) ion uptake from the aqueous solution at optimal conditions onto unloaded (a) and TAR-loaded PUFs sorbents (b) at 25°C after 1-h shaking.

where C_{ads} is the amount of mercury(II) ions retained per mass unit of PUFs, K_{DR} is the maximum amount of mercury(II) ions retained, β is a constant related to the energy of transfer of the solute from the bulk solution to the solid sorbent, and Є is Polanyi potential given by

$$\varepsilon^2 = RT\ln(1 + 1/Ce) \tag{18}$$

where R is gas constant in $KJmol^{-1}\ K^{-1}$ and T is the absolute temperature in Kelvin. The plot of ln C_{ads} *versus* $Є^2$ was linear with $R^2 = 0.996$ (**Figure 20**) for unloaded and TAR-loaded PUFs, meaning that D-R isotherm model is obeyed for mercury(II) ion sorption onto the sorbents over the entire concentration under study. The values of β and K_{DR} computed from the slope and intercept were equal to 0.008 $mol^2\ KJ^{-2}$, 4.48 mmol g^{-1} for TAR-loaded PUFs and 0.004 $mol^2\ KJ^{-2}$, 2.00 mmol g^{-1} for unloaded PUFs.

Assuming that the surface of PUF is heterogonous and an approximation to the Langmuir isotherm model is chosen as a local isotherm for all sites that are energetically equivalent, the quantity β can be related to the mean of free energy (E) of transferring one mole of the solute from infinity to the surface of PUFs. This quantity can be expressed by the following equation:

$$E = \frac{1}{\sqrt{-2\beta}} \tag{19}$$

The value of E was found to be 7.90 ± 0.07 KJ/mol for TAR-loaded PUFs sorbents.

With reference to the data reported previously [25], a dual sorption (involving absorption related to "solvent extraction" and an added component for "surface adsorption") is the most probable retention mechanism for mercury(II) ion uptake by the TAR-loaded foams. Such a proposed model can be written as follows:

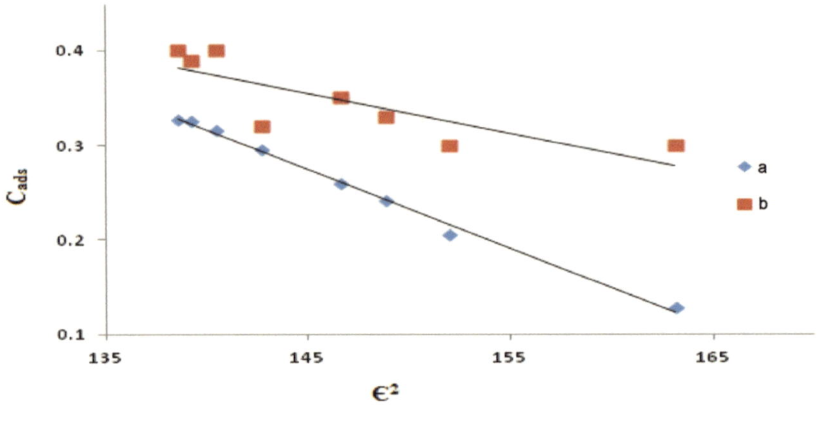

Figure 20. Dubinin-radushkevich (D-R) sorption isotherms of the mercury(II) ion uptake at optimal conditions by unloaded (a) and TAR-loaded PUFs sorbents (b) at 25°C after 1-h shaking.

$$C_r = C_{abs} + C_{ads} = DC_{aq} + \frac{SKLCaq}{1 + KLCaq} \tag{20}$$

where C_r and C_{aq} are the equilibrium concentrations of mercury(II) ions onto the solid sorbent and in the aqueous media, respectively. The parameters C_{abs} and C_{ads} are the equilibrium concentrations of mercury(II) ions onto the tested solid sorbents as absorbed and adsorbed species, respectively. S and K_L are the saturation values for the Langmuir adsorption.

3.5. Chromatographic separation of mercury(II) ions onto TAR-immobilized PUFs packed columns

The cellular membrane structures, excellent hydrodynamic and aerodynamic properties of PUFs [23, 25] enhanced the mercury(II) ion uptake onto TAR-treated PUFs packed columns. **Table 2** shows acceptable extraction and recovery percentages of mercury(II) ions in the range 88.1–103.7% (when percolating the aqueous media spiked with mercury(II) ions through TAR-treated foams packed columns and using HNO_3 as a proper eluting agent).

3.6. Analytical performance of the TAR-immobilized foam packed columns

The performance of TAR-treated PUF packed columns (0.40 ± 0.01 g) was determined by percolating a 2-L aqueous solution containing mercury(II) ions (20 μg mL^{-1}) through the packed columns at the optimal conditions (**Figure 21**). The values of HETP and N were calculated from the chromatogram by using Gluenkauf Eq. [17]:

$$N = \frac{8\,V^2_{max}}{W^2} = \frac{L}{HETP} \tag{21}$$

where V_{max} is the peak elution volume, W is the peak width at 1/e times the maximum solute concentration and L is the length of the foam bed in mm. The N and HETP values were found to be 89.00 ± 0.02 and 1.01 ± 0.02 mm, respectively.

The performance of TAR-treated PUF packed columns was also determined from the breakthrough capacity method i.e. percolating 2-L aqueous solution containing mercury(II)

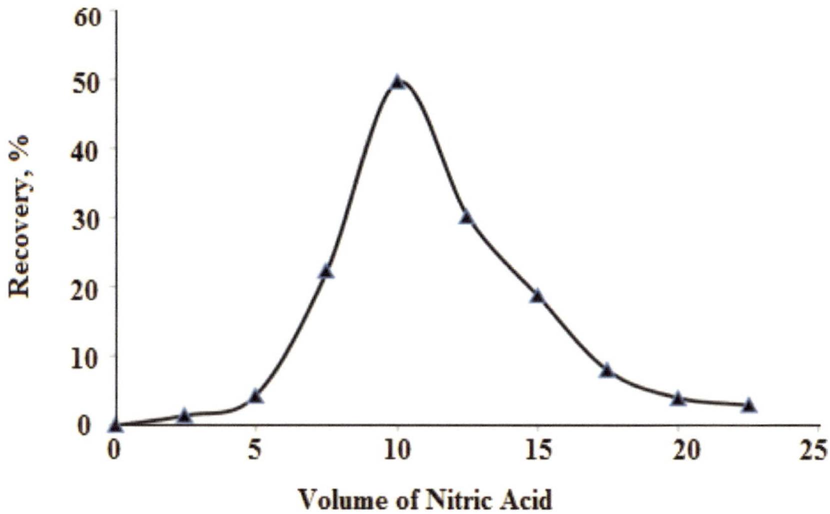

Figure 21. Chromatogram of mercury(II) ion extraction and recovery from the tested TAR- immobilized PUFs packed column using HNO_3 (1 Mol L^{-1}) as eluting agent at 5 mL min^{-1} flow rate.

ions (5 μg mL^{-1}) through the TAR loaded PUFs packed column at the optimal conditions (**Figure 22**). The values of HETP and N were then determined by using the Eq. [17]:

$$N = \frac{V_{50}\, V'}{(V_{50} - V')^2} = \frac{L}{HETP} \tag{22}$$

where V_{50} is the effluent volume at 50% breakthrough and V` is the volume that the column has a removal efficiency of 15.78% at the initial concentration. The values of N and HETP calculated from **Figure 22** were 90.00 ± 0.03 and 0.99 ± 0.02 mm respectively, and had good agreement with data obtained from the chromatogram method. The breakthrough capacity (BC) was also calculated by using the following equation:

$$BC = \frac{V_{50} \text{ x Co}}{\text{w}} \tag{23}$$

the value of BC was found to be 1.27 mg of mercury(II) ion uptake per one gram of the solid sorbent at 5 mL min^{-1} flow rate.

The sensitivity of this method was accessed by limit of detection (LOD) and limit of quantification (LOQ) under the established conditions for mercury(II) ion uptake by TAR- immobilized PUFs packed columns, using the following Equations [57].

$$LOD = 3\,S/b \tag{24}$$

and

$$LOQ = 10\,S/b \tag{25}$$

where *S* is the standard deviation (n = 5) of the blank and *b* is the slope of the calibration plot. The values of LOD and LOQ were found to be 0.024 and 0.080 ppm, respectively. Such limits

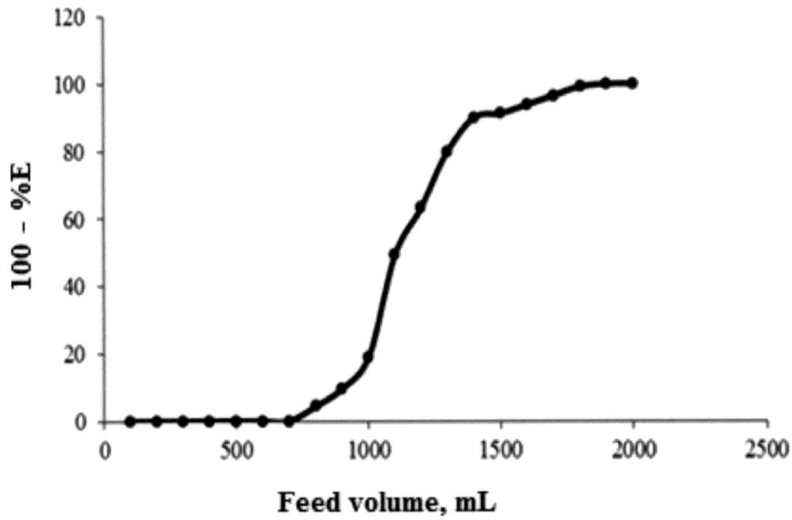

Figure 22. Breakthrough capacity curve for the mercury(II) ion retention onto the investigated columns at optimal conditions.

Hg(II) taken (μg L^{-1})	Hg(II) found (μg L^{-1})	Recovery (%)
500	518.4 ± 1.4	103.7 ± 0.8
100	102.8 ± 0.1	102.6 ± 1.3
10	9.9 ± 1.2	99.6 ± 1.1
5	4.4 ± 1.3	88.1 ± 1.5

Table 2. Analytical data of the extraction and recovery of mercury(II) ions by percolating 1-L deionized water spiked with mercury(II) ions through the tested TAR-PUFs packed columns at 5 mL min^{-1} flow rate (mean ± SD, n = 5).

Type of water	Hg(II) taken, ng mL^{-1}	Hg(II) found, ng mL^{-1}	Recovery %
Tap water	No	0.01	
	1	1.1	110.2 ± 0.1
	10	9.5	95.1 ± 0.1
	1000	825.9	82.5 ± 0.1
Sea water	No	0.03	
	1	0.93	93.6 ± 0.1
	10	7.4	74.5 ± 0.2
	1000	527.9	52.7 ± 0.1

Table 3. Analytical data for the determination of mercury(II) ions (spiked in tap and seawater samples) being removed by the proposed TAR-treated PUFs packed columns (mean ± SD, n = 5).

could be further lowered by increasing the sample volume of the aqueous phase containing ultra-traces of mercury(II) ions at the optimal experimental conditions.

3.7. Analytical applications of the developed TAR-PUFs packed columns

The accuracy of TAR-loaded PUFs packed columns was successfully evaluated by performing recovery tests for 1-L tap and sea water samples spiked with mercury(II) ions in the range 1–1000 ng mL^{-1}. These column proved to satisfactorily remove mercury(II) ions from the tested solutions (>52.7% as displayed in **Table 3**).

4. Conclusion

The kinetic and thermodynamic characteristics of mercury(II) ion uptake by 4-(2-thiazolyl azo) resorcinol-immobilized PUFs sorbents were critically investigated. The kinetic data confirmed the intra-particle diffusion and first-order kinetics of mercury(II) ion retention onto the tested sorbents. Exothermic nature of the retention process of mercury(II) ion onto TAR-treated PUFs is governed by the negative value of ΔH. On the other hand, the negative value of ΔG proves that mercury(II) ion sorption onto the tested sorbents is a spontaneous phenomenon. The developed PUFs packed columns provide a simple, reliable, fairly rapid and low-cost method for the pre-concentration, separation and determination of mercury(II) ion in the aqueous media.

Acknowledgements

The authors would like to thank King Abdulaziz City for Sciences and Technology (KACST), Saudi Arabia for the fund provided to one of the coauthors (A.H. Al-Bagawi). Special thanks to the Deanship of Scientific Research (DSR) at King Abdulaziz University, Jeddah for the facilities and technical support.

Author details

Amal H. Al-Bagawi[1,2], Waqas Ahmad[1], Hassan Alwael[1], Zeinab M. Saigl[1], Gharam I. Mohammed[3], Yousry M. Moustafa[3], Eman A. Al-Harbi[4] and Mohammad S. El-Shahawi[1]*

*Address all correspondence to: mohammad_el_shahawi@yahoo.co.uk

1 Department of Chemistry, Faculty of Science, King Abdulaziz University, Jeddah, Saudi Arabia

2 Department of Chemistry, Faculty of Science, Hail University, Hail, Saudi Arabia

3 Department of Chemistry and Physics, Faculty of Applied Science, Umm Al-Qura University, Makkah, Saudi Arabia

4 Department of Chemistry, Faculty of Science, Taibah University, Al-Madina Al-Mounawara, Saudi Arabia

References

[1] Davis ML, Cornwell DA, editors. Introduction to Environmental Engineering. 5th ed. New York: McGraw-Hill; 2012. 1024 p

[2] Miretzky P, Cirelli AF. Hg(II) removal from water by chitosan and chitosan derivatives: A review. Journal of Hazardous Materials. 2009;**167**(1-3):10-23

[3] World Health Organization. Guidelines for Drinking Water Quality. 4th edition. incorporating the 1st addendum. 2017. pp. 389-390

[4] Edzwald JK, editor. Water Quality and Treatment: A Handbook on Drinking Water. 6th ed. New York: American Water Works Association, American Society of Civil Engineers, McGraw Hill Inc.; 2011. 1969 p

[5] National Research Council, (US) Committee on the Toxicological Effects of Methylmercury. Toxicological Effects of Methylmercury. Washington, DC: National Academies Press; 2000. 343 p

[6] Osmond DL, Gannon RW, Gale JA, Line DE, Knott CB, Phillips KA, et al. A decision support system for watershed-scale nonpoint source water quality problems. Journal of the American Water Resources Association. 1997;**33**:327-341

[7] Syversen T, Kaur P. The toxicology of mercury and its compounds. Journal of Trace Elements in Medicine and Biology. 2012;**26**(4):215-226

[8] Banaee M, Beitsayah A, Jorabdoz I. Assessment of mercury bioaccumulation in Zebra cichlid (*Cichlasoma nigrofasciatum*) exposed to sublethal concentrations of permethrin. Iranian Journal of Toxicology. 2015;**8**(27):1168-1173

[9] Armenta S, Garrigues S, De la Guardia M. Green analytical chemistry. Trends in Analytical Chemistry. 2008;**27**(6):497-511

[10] Babaei A, Shams E, Samadzadeh A. Simultaneous determination of copper, bismuth and lead by adsorptive stripping voltammetry in the presence of thymolphthalexone. Analytical Sciences. 2006;**22**(7):955-959

[11] Bloom N. Determination of picogram levels of methylmercury by aqueous phase ethylation, followed by cryogenic gas chromatography with cold vapour atomic fluorescence detection. Canadian Journal of Fisheries and Aquatic Sciences. 1989;**46**(7):1131-1140

[12] Capitán-Vallvey LF, Navas Iglesias N, Orbe Paya ID, Avidad Castañeda R. Simultaneous determination of quinoline yellow and brilliant blue FCF in cosmetics by solid-phase spectrophotometry. Talanta. 1996;**43**(9):1457-1464

[13] Chatterjee S, Pillai A, Gupta VK. Spectrophotometric determination of mercury in environmental sample and fungicides based on its complex with o-carboxy phenyl diazoamino p-azobenzene. Talanta. 2002;**57**(3):461-465

[14] Bennun L, Gomez J. Determination of mercury by total-reflection X-ray fluorescence using amalgamation with gold. Spectrochimica Acta Part B: Atomic Spectroscopy. 1997;**52**(8):1195-1200

[15] Al-Bagawi AH, Ahmad W, Saigl ZM, Alwael H, Al-Harbi EA, El-Shahawi MS. A simple and low cost dual-wavelength β-correction spectrophotometric determination and speciation of mercury (II) in water using chromogenic reagent 4-(2-thiazolylazo) resorcinol. Spectrochimica Acta Part A: Molecular and Biomolecular Spectroscopy. 2017;**187**:174-180

[16] Moody GJ, editor. Chromatographic Separation with Foamed Plastics and Rubbers (Chromatographic Science Series). 1st ed. Vol. 152. New York: Marcel Dekker Inc; 1982

[17] Braun T, Navratil JD, Farag AB, editors. Polyurethane Foam Sorbents in Separation Science. Boca Rotan, Florida: CRC Press Inc.; 1985. 219 p

[18] Alfassi ZB, Wai CM, editors. Preconcentration techniques for trace elements. Boca Rotan, Florida: CRC Press Inc.; 1991. 480 p

[19] El-Shahawi MS, Bashammakh AS, Abdelmageed M. Chemical speciation of chromium (III) and (VI) using phosphonium cation impregnated polyurethane foams prior to their spectrometric determination. Analytical Sciences. 2011;**27**(7):757-763

[20] El-Shahawi MS, Al-Saidi HM, Bashammakh AS, Al-Sibaai AA, Abdelfadeel MA. Spectrofluorometric determination and chemical speciation of trace concentrations of chromium (III & VI) species in water using the ion pairing reagent tetraphenyl-phosphonium bromide. Talanta. 2011;**84**(1):175-179

[21] El-Shahawi MS, Alwael H, Arafat A, Al-Sibaai AA, Bashammakh AS, Al-Harbi EA. Kinetics and thermodynamic characteristics of cadmium (II) sorption from water using procaine hydrochloride physically impregnated polyurethane foam. Journal of Industrial and Engineering Chemistry. 2015;**28**:147-152

[22] Farag AB, Soliman MH, Abdel-Rasoul OS, El-Shahawi MS. Sorption characteristics and chromatographic separation of gold (I and III) from silver and base metal ions using polyurethane foams. Analytica Chimica Acta. 2007;**601**(2):218-229

[23] El-Shahawi MS, Al-Sibaai AA, Bashammakh AS, Alwael H, Al-Saidi HM. Ion pairing based polyurethane foam sorbent packed column combined with inductively coupled plasma–optical emission spectrometry for sensitive determination and chemical speciation of bismuth (III & V) in water. Journal of Industrial and Engineering Chemistry. 2015;**28**:377-383

[24] Oribayo O, Feng X, Rempel GL, Pan Q. Synthesis of lignin-based polyurethane/graphene oxide foam and its application as an absorbent for oil spill clean-ups and recovery. Chemical Engineering Journal. 2017;**323**:191-202

[25] El-Shahawi MS, Nassif HA. Kinetics and retention characteristics of some nitrophenols onto polyurethane foams. Analytica Chimica Acta. 2003;**487**(2):249-259

[26] El-Shahawi MS, El-Sonbati MA. Retention profile, kinetics and sequential determination of selenium (IV) and (VI) employing 4,4′-dichlorodithizone immobilized-polyurethane foams. Talanta. 2005;**67**(4):806-815

[27] Han J, Cao Z, Qiu W, Gao W, Hu J, Xing B. Probing the specificity of polyurethane foam as a 'solid-phase extractant': Extractability-governing molecular attributes of lipophilic phenolic compounds. Talanta. 2017;**172**:186-198

[28] El-Shahawi MS, Al-Sibaai AA, Bashammakh AS, Alwael H, Al-Saidi HM. Ion pairing based polyurethane foam sorbent packed column combined with inductively coupled plasma–optical emission spectrometry for sensitive determination and chemical speciation of bismuth(III & V) in water. Journal of Industrial and Engineering Chemistry. 2015;**28**:147-152

[29] Choi H, Woo NC, Jang M, Cannon FS, Snyder SA. Magnesium oxide impregnated polyurethane to remove high levels of manganese cations from water. Separation and Purification Technology. 2014;**136**:184-189

[30] Liu Y, Ma J, Wu T, Wang X, Huang G, Liu Y, et al. Cost-effective reduced graphene oxide-coated polyurethane sponge as a highly efficient and reusable oil-absorbent. ACS Applied Materials & Interfaces. 2013;**5**(20):10018-10026

[31] Gu R, Sain MM, Konar SK. A feasibility study of polyurethane composite foam with added hardwood pulp. Industrial Crops and Products. 2013;**42**:273-279

[32] Wang Q, Wang H, Xiong S, Chen R, Wang Y. Extremely efficient and recyclable absorbents for oily pollutants enabled by ultrathin-layered functionalization. ACS Applied Materials & Interfaces. 2014;**6**(21):18816-18823

[33] Nikkhah AA, Zilouei H, Asadinezhad A, Keshavarz A. Removal of oil from water using polyurethane foam modified with nanoclay. Chemical Engineering Journal. 2015; **262**:278-285

[34] Wang G, Zeng Z, Wu X, Ren T, Han J, Xue Q. Three-dimensional structured sponge with high oil wettability for the clean-up of oil contaminations and separation of oil–water mixtures. Polymer Chemistry. 2014;**5**(20):5942-5948

[35] Cinelli P, Anguillesi I, Lazzeri A. Green synthesis of flexible polyurethane foams from liquefied lignin. European Polymer Journal. 2013;**49**(6):1174-1184

[36] Abu-Zahra N, Gunashekar S. Structurally functionalized polyurethane foam for elimination of lead ions from drinking water. Journal of Research Updates in Polymer Science. 2014;**3**(1):16-25

[37] Moawed EA, Farag AB, El-Shahat MF. Separation and determination of some trivalent metal ions using rhodamine B grafted polyurethane foam. Journal of Saudi Chemical Society. 2013;**17**(1):47-52

[38] Mangaleshwaran L, Thirulogachandar A, Rajasekar V, Muthukumaran C, Rasappan K. Batch and fixed bed column studies on nickel (II) adsorption from aqueous solution by treated polyurethane foam. Journal of the Taiwan Institute of Chemical Engineers. 2015; **55**:112-118

http://dx.doi.org/10.5772/intechopen.81206

[39] Góes MM, Keller M, Oliveira VM, Villalobos LD, Moraes JC, Carvalho GM. Polyurethane foams synthesized from cellulose-based wastes: Kinetics studies of dye adsorption. Industrial Crops and Products. 2016;**85**:149-158

[40] Barbara I, Dourges MA, Deleuze H. Preparation of porous polyurethanes by emulsion-templated step growth polymerization. Polymer. 2017;**132**:243-251

[41] Pérez MC, Álvarez-Hornos FJ, Engesser KH, Dobslaw D, Gabaldón C. Removal of 2-butoxyethanol gaseous emissions by biotrickling filtration packed with polyurethane foam. New Biotechnology. 2016;**33**(2):263-272

[42] Khulbe KC, Matsuura T. Removal of heavy metals and pollutants by membrane adsorption techniques. Applied Water Science. 2018;**8**(1):19

[43] Gunashekar S, Abu-Zahra N. Synthesis of functionalized polyurethane foam using BES chain extender for lead ion removal from aqueous solutions. Journal of Cellular Plastics. 2015;**51**(5-6):453-470

[44] Hussein FB, Abu-Zahra NH. Synthesis, characterization and performance of polyurethane foam nanocomposite for arsenic removal from drinking water. Journal of Water Process Engineering. 2016;**13**:1-5

[45] Hussein FH, Abu-Zahra NH. Extended performance analysis of polyurethane-iron oxide nanocomposite for efficient removal of arsenic species from water. Water Science & Technology: Water Supply. 2017;**17**(3):889-896

[46] El-Shahawi MS, Hamza A, Al-Sibaai AA, Al-Saidi HM. Fast and selective removal of trace concentrations of bismuth (III) from water onto procaine hydrochloride loaded polyurethane foams sorbent: Kinetics and thermodynamics of bismuth (III) study. Chemical Engineering Journal. 2011;**173**(1):29-35

[47] Chung YS, Chung WS. Determination of Co (II) ion as a 4-(2-thiazolylazo) resorcinol or 5-methyl-4-(2-thiazolylazo) resorcinol chelate by reversed-phase capillary high-performance liquid chromatography. Bulletin of the Korean Chemical Society. 2003;**24**(12):1781-1784

[48] Svehla G, editor. Vogel's Quantitative Inorganic Analysis. 7th ed. Prentice Hall; 1996. 347 p

[49] Weber WJ, Morris JC. Kinetics of adsorption on carbon from solution. Journal of the Sanitary Engineering Division Proceedings. 1963;**89**(2):31-60

[50] Lagergren S, Sven BK. About the theory of so-called adsorption of soluble substances. Kungliga Svenska Vetenskapsakademiens Handlingar. 1898;**24**(4):1-39

[51] Bhattacharya AK, Enkobachar CV. Removal of cadmium (ii) by low cost adsorbents. Journal of Environmental Engineering. 1984;**110**(1):110-122

[52] Bashammakh AS. The retention profile of phosphate ions in aqueous media onto ion pairing immobilized polyurethane foam: Kinetics, sorption and chromatographic separation. Journal of Molecular Liquids. 2016;**220**:426-431

[53] Ho YS, McKay G. Pseudo-second order model for sorption processes. Process Biochemistry. 1999;**34**(5):451-465

[54] Freundlich H. Capillary and Colloid Chemistry. London: Methuen and Co. Ltd; 1926. pp. xv+883

[55] Langmuir I. The adsorption of gases on plane surfaces of glass, mica and platinum. Journal of the American Chemical Society. 1918;**40**(9):1361-1403

[56] Somorjai GA, Li Y. Introduction to Surface Chemistry and Catalysis. 2nd ed. New York: John Wiley & Sons, Inc.; 2010. p. 800

[57] Miller JC, Miller JN. Statistics and Chemometrics for Analytical Chemistry. 6th ed. Canada: Pearson Education; 2010. 296 p

Chapter 4

Variability Characterization of the Olive Species Regarding Virgin Olive Oil Aroma Compounds by Multivariate Analysis of GC Data

Carlos Sanz, Angjelina Belaj, Mar Pascual and
Ana G. Pérez

Additional information is available at the end of the chapter

http://dx.doi.org/10.5772/intechopen.76343

Abstract

Virgin olive oil is characterized by its unique aroma, which is synthesized when olive fruits are crushed during the industrial process used for oil production. The genetic variability of the major volatile compounds comprising the oil aroma was studied in a representative sample of olive cultivars from the World Olive Germplasm Collection (IFAPA, Cordoba, Spain). The analytical data demonstrated that a high degree of variability for the content of volatile compounds is found in the olive species and that most of the volatile compounds found in the oils were synthetized by the enzymes included in the so-called lipoxygenase pathway. The use of multivariate analysis to identify cultivars is particularly interesting in terms of volatile composition and deduced organoleptic quality. It can be used for identification of old olive cultivars that give rise to oils with a high organoleptic quality and in parent selection for olive breeding programs.

Keywords: *Olea europaea* L., virgin olive oil, volatile compounds, variability, quality

1. Introduction

Virgin olive oil (VOO) is an essential component of the traditional Mediterranean diet that attracts the interest of the scientific community for its health-promoting properties. This diet is associated with a lower incidence of a number of diseases related to inflammatory processes such as cardiovascular diseases, diabetes, metabolic syndrome, arthritis, Alzheimer's disease and certain types of cancer [1–6]. Among other plant oils, VOO is unique for its high levels of monounsaturated fatty acids and the presence of a wide range of minor components that are largely responsible for both their health-promoting properties and organoleptic characteristics. As for the organoleptic characteristics, phenolic compounds are closely related to the pungent and bitter sensory notes while volatile minor components are consequently responsible for the aroma of VOO. Curiously, a possible interacting effect of phenolic compounds on the VOO aroma release and perception has been recently reported [7].

About 25 years ago, we established that most of the volatile compounds present in VOO are synthesized when enzymes and substrates come together during the olive fruit milling in the oil extraction process, and that the enzymatic activities within the lipoxygenase (LOX) pathway are involved in this synthesis [8]. Aldehydes and alcohols of six straight-chain carbons (C6), as well as their corresponding esters, are the most important compounds in VOO aroma, either qualitatively or quantitatively [9, 10]. Linoleic (LA) and linolenic (LnA) acids are the main substrates for this synthesis. As displayed in **Figure 1**, LOX enzymes catalyze the production of 13-hydroperoxide derivatives in the first step of this pathway, which are subsequently cleaved heterolytically by hydroperoxide lyase (HPL) enzyme to form C6 aldehydes [8, 11, 12]. C6 aldehydes are then reduced by alcohol dehydrogenase (ADH) enzymes to C6 alcohols [8, 13], and they are finally transformed into the corresponding esters by alcohol acyltransferase (AAT) enzymes [8, 14]. Angerosa et al. [10] also demonstrated the relevance of five straight-chain carbon (C5) compounds in the aroma of olive oil. They are supposedly generated through the production of a 13-alcoxyl radical in an additional branch of the LOX pathway, as demonstrated in soybean seeds [15]. This 13-alcoxyl radical would be homolytically cleaved to form a 1,3-pentene allylic radical. The latter can be chemically dimerized to generate pentene dimers (PD) or reacts with hydroxyl radicals to form C5 alcohols, which are in turn oxidized to C5 carbonyl compounds by ADH enzyme activities, as believed to occur also in soybean [16]. The synthesis of these volatile compounds seems to depend mainly on the availability of fatty acid substrates to be catabolized through the LOX pathway during the VOO extraction process. In addition, LOX activity also proved to be an important limiting factor for the synthesis of these volatile compounds, and this limitation is seemingly characteristic of each olive cultivar [17, 18].

One of the key characteristics of the olive species is its wide genetic patrimony. The establishment of *ex-situ* germplasm banks has received much attention for the collection and conservation of olive genetic resources. In this sense, the World Olive Germplasm Collection (WOGC, Cordoba, Spain) is an international reference due to a high number of cultivars included in its database as well as a high degree of evaluation and identification of cultivars by molecular markers and agronomical traits [19]. This high genetic diversity of the olive germplasm collection could be very useful to recuperate old cultivars, which produce oils with outstanding aromas, or for the selection of optimal parents for olive breeding programs with the aim of finding new cultivars with improved olive oil aroma. New olive cultivars with improved organoleptic quality might further stimulate VOO consumption. In this sense, olive breeding programs have been lately addressing the selection for sensory and nutritional quality of VOO [20, 21] in addition to the traditional improvement of agronomic traits [22].

http://dx.doi.org/10.5772/intechopen.76343

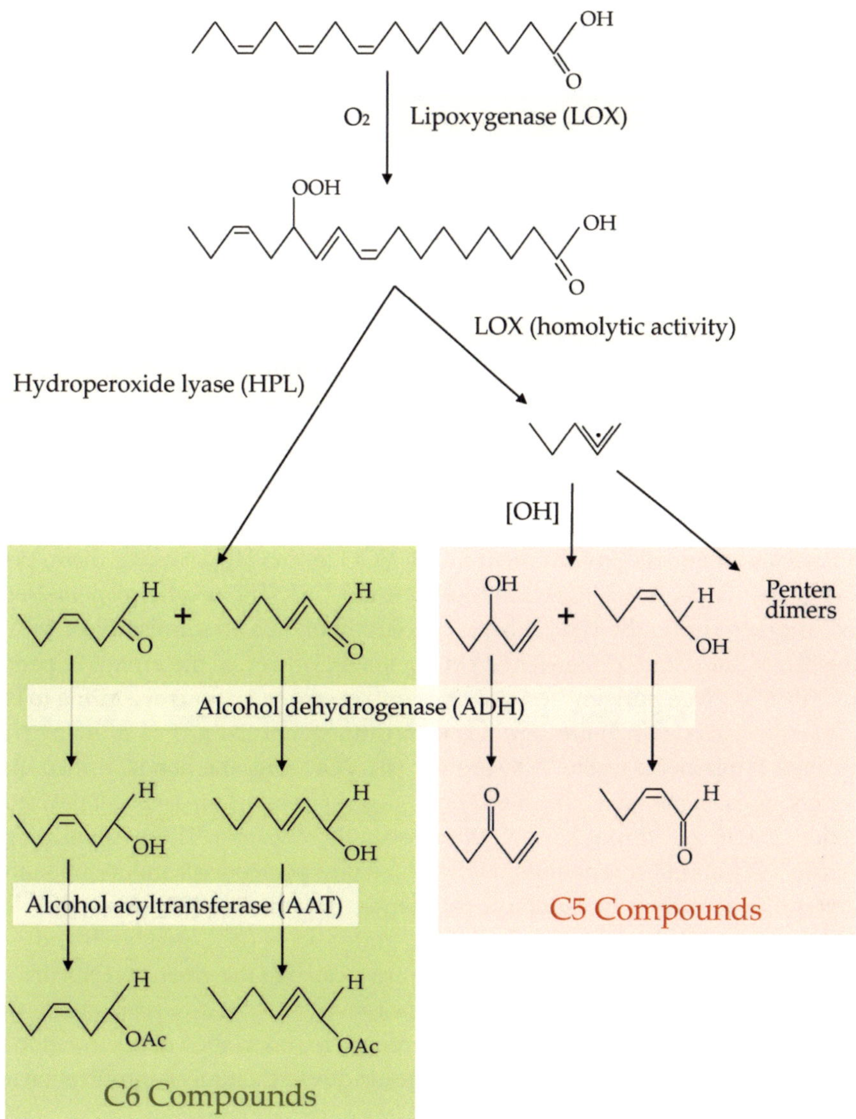

Figure 1. Lipoxygenase pathway on linolenic acid.

2. Sensory properties

In addition to its possible health benefits, the excellent organoleptic properties may well explain the continued increase in the demand for high-quality VOO [23]. As mentioned earlier, volatile compounds are responsible for the aroma of this product, which is basically composed, when obtained from sound fruits, of a mixture of green and fruity odor notes spiced with some other positive notes, making it a unique edible oil. Odor perception and pleasantness are determined by the size, shape, conformation, type and position of the functional groups in the volatile compound [24, 25]. In this sense, Aparicio and Morales [26] developed a statistical sensory wheel (SSW) to understand the relationship between volatile compounds and odor attributes in VOO, through a compilation of the data obtained from trained VOO sensory panels across Europe. As a result, odor notes with a similar semantic description are clustered into sectors together with volatile compounds identified by a given sensory perception. Among the most relevant to VOO aroma, this sensory wheel includes the green or ripe fruit odor notes. Moreover, the sensory attribute of each volatile compound depends on its concentration and odor threshold in the oil. The ratio between the concentration of the volatile compound and its odor threshold is the odor activity value (i.e. OAV). Those volatile compounds with an OAV below one do not theoretically contribute to the aroma of VOO.

Among the components of the volatile fraction of VOO, C6 compounds derived from LnA (C6/LnA) are considered the main contributors to its aroma. However, those compounds derived from LA (C6/LA) provide unpleasant sensations according to literature, especially the hexan-1-ol [10, 26, 27]. As mentioned earlier, C5 compounds derived from LnA (C5/LnA) seem to have also a notable involvement in the VOO aroma [10]. Among them, pent-1-en-3-one and pent-2-en-1-ols are especially noteworthy [28]. The aroma of pent-2-en-1-ol is described as green-fruity, the typical basic perception of VOO, reminiscent of healthy and fresh olive fruits harvested at the right ripening stage. However, the aroma of pent-1-en-3-one is described as green pungent and provides unpleasant sensations according to literature [10, 26]. Volatile esters, which are significant constituents of the aroma of many fruits, are also important components of the VOO aroma. The VOO ester fraction is formed mainly by acetate esters, such as the hexyl, (*E*)-hex-2-en-1-yl and (*Z*)-hex-3-en-1-yl acetates, from alcohol moieties synthesized through the LOX pathway (LOX esters). These esters are grouped among those VOO aroma compounds with green-fruity perceptions and in the limits of the sweet perception according to canonical correlations reported in [28]. Also, the aldehydes derived from the metabolism of branch-chained amino acids (BC), such as 2- and 3-methylbutanal, provide green-fruity odor notes as they are located in the green and ripe fruit sectors of the SSW [26]. However, the alcohol derivative 2-methyl-butan-1-ol seems to be related to the fusty defects of VOO aroma [27]. Finally, terpenes found in the volatile fraction of VOO seem not to be important contributors to VOO aroma due to their low concentration and high odor threshold.

The volatile fraction of VOO is also responsible for the off-flavors present in some oils. This is very important because VOO is the only food product that requires a sensory analysis to be classified into commercial categories. This classification is ruled by the European Official Regulations for olive oil [29] and carried out by trained test panels, in which the evaluation of aroma defects plays a very important role. When VOO was obtained from infested olive fruits or fruits picked from the ground, or if VOO was inadequately processed or stored, the volatile fraction of VOO is altered and may include compounds that are responsible for off-flavors. They are predominantly carboxylic acids or aliphatic C8-C11 carbonyls and alcohols [30, 31] produced by chemical oxidation or through the activity of exogenous enzymes, usually due to microbial activity present in non-sound fruits. The presence of such compounds is commonly associated with sensory defects in the oils.

3. Natural variation of aroma compounds

As mentioned earlier, the extensive genetic patrimony of the olive tree is one of the main attributes of this crop, which is represented by thousands of cultivars from specific areas in the Mediterranean basin where they were originally cultivated. This genetic heritage is currently used in different olive breeding programs, but it is only that the sensory properties of VOO are considered as main traits, in addition to the agronomic traits [20, 22, 32]. Regarding these sensory properties, the natural variation of aroma compounds in VOO was studied using the WOGC olive cultivar collection. For this purpose, a quarter of this germplasm collection (**Table 1**) was considered. This subset included cultivars from the Core-36 collection from WOGC, which preserves most of the genetic diversity found in this germplasm collection [19]. To better compare the different olive accessions, the olive trees from this collection were cultivated in the same edaphoclimatic conditions and only healthy fruits were picked at the turning stage during a period of 3 years. In addition, mild operating conditions were employed for the oil extraction process to minimize the production of compounds that cause sensory defects in the oils.

Solid-phase microextraction (SPME) of the head space (HS) in combination with gas chromatography and mass spectrometry (GC–MS) is a very powerful technique that is used quite frequently for the analysis of volatile compounds in foods. HS-SPME–GG–MS has been applied to study the aroma of products derived from the olive fruit, such as oil or table olives [18, 33–35]. Actually, SPME-GC–MS is the method of reference to validate the discriminating power of new e-sensing technologies such as the electronic nose for olive oil [36–38]. Analysis of volatile compounds in VOO from the WOGC olive cultivar collection was performed by HS-SPME-GC–MS according to Pérez et al. [33]. Olive oil samples (0.5 g) were prepared in 10-mL vials and were conditioned to room temperature before being placed in a vial heater at 40°C for a 10-min equilibration time. Volatile compounds from the headspace were adsorbed onto SPME fiber DVB/Carboxen/PDMS 50/30 μm (Supelco Co., Bellefonte, PA). Sampling time was carried out at 40°C for 50 min. Desorption of volatile compounds was completed directly into the GC injector. Volatile compounds were identified out on a 7820A/GC-5975/MSD system (Agilent Technologies), equipped with a DB-Wax capillary column (60 m × 0.25 mm i.e., film thickness, 0.25 μm: J&W Scientific, Folsom, CA) and under the following conditions: the injection port was operated in splitless mode at 250°C; He was used as the carrier gas and the flow rate was 1 mL/min; the column was held for 6 min at 40°C and then programmed at 2°C min^{-1} to 168°C; the mass detector was operated in the electronic impact mode at 70 eV, the source temperature was set at 230°C and the mass spectra were scanned at 2.86 scans/s in the m/z 40–550 amu range. Compound identification was performed by matching against the Wiley/NBS Library and by comparing GC retention time against standards. For quantitative purposes, the volatile fraction was analyzed three times on a HP-6890 GC apparatus (Agilent Technologies, Santa Clara, CA, USA), which was equipped with a similar column and operated under the following operating conditions in order to reproduce the same retention times for volatile compounds as those obtained with the 7820A/GC-5975/MSD system: N_2 was used as the carrier gas at a constant pressure of 17 psi; the injector and detector were maintained at 250°C; the column was held for 6 min at 40°C and then programmed at 2°C min^{-1} to 168°C. For quantification, calibration curves for each compound were made in re-deodorized high-oleic sunflower oil.

No.	Cultivar	No.	Cultivar
1	Picudo	50	Sabatera
2	Lechín de Sevilla	51	Arbosana
3	Picual	52	Joanenca
4	Cornicabra	53	Menya
5	Blanqueta	54	Elmacik
6	Manzanilla de Sevilla	55	Kan Celebi
7	Jaropo	56	Yun Gelebi
8	Blanqueta-48	57	Kalokerida
9	Verdial de Vélez-Málaga	58	Agouromanakolia
10	Picholine	59	Amygdalolia Nana
11	Caninese	60	Mavreya
12	Coratina	61	Myrtolia
13	Frantoio	62	Kolybada
14	Leccino	63	Levantinka
15	Maurino	64	Lastovka
16	Cipresino	65	Barnea
17	Uslu	66	Aggezi Shami
18	Picholine Marroquí	67	Wardan
19	Kalamon	68	Azapa o Arauco
20	Megaritiki	69	Istarska Bjelica
21	Ouslati	70	Manzanillera de Huercal Overa
22	Zalmati	71	Torcio de Cabra
23	Chemial de Kabilye	72	Zaity
24	Merhavia	73	Figueretes
25	Galega Vulgar	74	Haouzia
26	Kaesi	75	Abou Kanani
27	Majhol-152	76	Abbadi Abou Gabra
28	Grappolo	77	Verdial de Badajoz
29	Koroneiki	78	Verdial de Vélez-Málaga
30	Morrut	79	Verde Verdelho
31	Llumeta	80	Piñonera
32	Rapasayo	81	Majhol-1013
33	Mastoidis	82	Barri

No.	Cultivar	No.	Cultivar
34	Temprano	83	Abbadi Abou Gabra
35	Lentisca	84	Shami
36	Chorreao de Montefrío	85	Abou Satl Mohazam
37	Pecoso	86	Vinyols
38	Negrinha	87	Argudell
39	Dokkar	88	Curivell
40	Pequeña de Casas Ibañez	89	Ulliri i Kuq
41	Borriolenca	90	Klon-1081
42	Vallesa	91	Arbequina
43	Patronet	92	Jabali
44	Forastera de Tortosa	93	Maarri
45	Canetera	94	Shengeh
46	Corbella	95	Mari
47	Vera	96	Bosana
48	Palomar	97	Klon-1812
49	Vaneta		

Table 1. Cultivars from the WOGC subset.

The subset of the olive germplasm collection showed a high degree of variability in terms of the content of volatile compounds (**Figure 2** and **Table 2**). As mentioned earlier, most of the volatile compounds found in the oils were synthetized by the enzymes included in the so-called LOX pathway [8], and can be grouped according to the length of the chain, the fatty acid substrate (C6/LnA, C6/LA, C5/LnA, C5/LA) and the origin of the esters (LOX esters). Moreover, a group of compounds derived from amino acid metabolism, which has a branched-chain chemical structure (BC) and a terpene also contributed quantitatively to the volatile fraction of VOO. C6 compounds derived from linolenic acid (C6/LnA) represented on average about 70% of total volatile fraction in the oils (**Figure 2**). The C6/LnA compounds varied from 5.18 to 67.94 μg/g oil, a variability that broadly exceeded that found for oils produced from a collection of 39 olive cultivars cultivated in the same orchard under the same edaphoclimatic conditions (2.52–18.11 μg/g oil) [39]. The C6/LnA aldehyde group was the most abundant, being (*E*)-hex-2-enal as the main compound (93% of total C6/LnA compounds on average, **Figure 2**). The mean content of (*E*)-hex-2-enal in the oils was 23.78 μg/g oil, ranging from 4.00 to 65.03 μg/g oil. The large amount and the relatively low odor threshold [40] point to this C6/LnA compound as one of the main contributors to the aroma of the oils, which seems to be a common feature of the oils from the olive species [33, 39, 41].

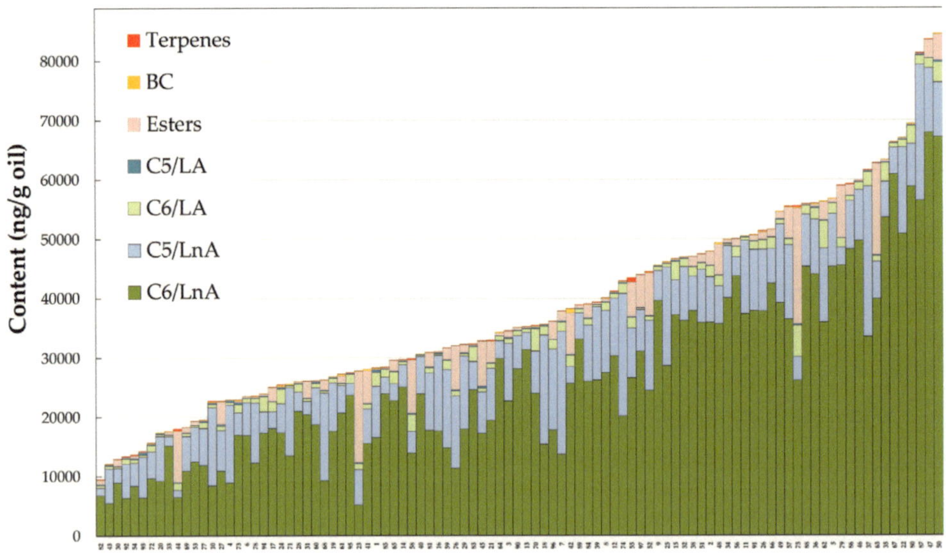

Figure 2. Content (ng/g oil) of the main groups of volatile compounds in the oils of WOGC cultivar subset. Numbers correspond to the WOGC cultivars displayed in **Table 1**.

Only the C5/LnA compounds showed contents close to those of the C6/LnA compounds. The pentene dimers are the most important compounds in this group from a quantitative point of view. They are thought to be produced through the same branch of the LOX pathway as the rest of C5 compounds [42]. The pentene dimers represented 82% of the C5/LnA content on average (**Figure 2**), but seem to have a quite low or negligible sensory contribution to the VOO aroma according to the estimated odor thresholds for these compounds [33]. The rest of C5/LnA compounds seem to contribute to VOO aroma according to their OAVs (**Table 2**). Pent-1-en-3-one and pent-2-en-1-ols are especially remarkable. All the cultivars showed to have contents of pent-1-en-3-one above its odor threshold (0.73 ng/g oil). The odor of this compound provides an unpleasant sensation [10], while the aroma of pent-2-en-1-ol is described as green-fruity. Most of the olive accessions (79%) presented (*E*)-pent-2-en-1-ol contents above its odor threshold [33] (**Table 2**). On the contrary, (*Z*)-pent-2-en-1-ol contents suggest this compound is of little relevance to VOO aroma as it is present below its threshold concentration in the oils.

As mentioned previously, fruity odor notes are considered as positive attributes of the oil aroma. Volatile esters are the main compounds responsible for these odor notes, especially LOX esters. The LOX ester content in the olive collection subset was 1953 ng/g oil on average and is present in a range of 14–19,334 ng/g oil, much higher than those values found in a progeny of the cross of cultivars Picual and Arbequina [33]. Only a few accessions had contents of (*E*)-hex-2-en-1-yl acetate contributing to VOO aroma (OAV > 1, **Table 2**). However, (*Z*)-hex-3-en-1-yl acetate did seem to be an important contributor to VOO aroma in more than half of the olive collection subset (**Table 2**). Similar results were found in the analysis of the volatile aroma compounds of some Turkish cultivar oils [43].

	Volatile compound	Code	Min	Max	Mean	% cv OAV > 1
C6/LnA aldehydes	(E)-hex-3-enal	6C-1	55	924	342	23
	(Z)-hex-3-enal	6C-2	52	4598	837	100
	(Z)-hex-2-enal	6C-3	101	1290	614	77
	(E)-hex-2-enal	6C-4	3998	65,029	23,780	100
C6/LnA alcohols	(E)-hex-3-enol	6C-5	0	836	32	0
	(Z)-hex-3-enol	6C-6	45	4306	586	12
	(E)-hex-2-enol	6C-7	43	3293	235	0
C6/LA aldehyde	hexanal	6C-8	166	4451	927	84
C6/LA alcohol	hexan-1-ol	6C-9	13	2725	339	25
C5/LnA carbonyls	pent-1-en-3-one	5C-1	68	1615	565	100
	(Z)-pent-2-enal	5C-2	4	175	47	0
	(E)-pent-2-enal	5C-3	22	226	93	0
C5/LnA alcohols	pent-1-en-3-ol	5C-4	29	634	228	7
	(Z)-pent-2-en-1-ol	5C-5	10	290	58	2
	(E)-pent-2-en-1-ol	5C-6	79	1908	480	79
Pentene dimers	pentene dimer - 1	5C-7	32	1195	406	0
	pentene dimer - 2	5C-8	26	1036	310	0
	pentene dimer - 3	5C-9	116	4840	1537	0
	pentene dimer - 4	5C-10	127	6363	1697	0
	pentene dimer - 5	5C-11	0	4513	1024	0
	pentene dimer - 6	5C-12	37	3909	1088	0
	pentene dimer - 7	5C-13	0	2847	571	0
C5/LA carbonyls	pentan-3-one	5C-14	17	335	89	0
	pentanal	5C-15	5	242	49	1
C5/LA alcohol	pentan-1-ol	5C-16	0	37	7	0
LOX esters	hexyl acetate	E-1	5	8165	670	15
	(E)-hex-2-en-1-yl acetate	E-2	7	2099	99	5
	(Z)-hex-3-en-1-yl acetate	E-3	0	11,120	1184	52
non-LOX esters	methyl acetate	E-4	5	42	18	0
	ethyl acetate	E-5	3	354	21	0
	methyl hexanoate	E-6	6	128	25	0
	ethyl hexanoate	E-7	0	359	37	0

	Volatile compound	Code	Min	Max	Mean	% cv OAV > 1
BC aldehydes	3-methyl-butanal	BC-1	6	351	40	100
	2-methyl-butanal	BC-2	5	342	25	100
BC alcohol	2-methyl-butan-1-ol	BC-3	5	159	34	0
Terpene	limonene	T-1	0	743	55	3

Table 2. Content of the volatile compounds (ng/g oil) and their distribution in the oils from the WOGC cultivar subset.

The levels of the branched-chain (BC) compounds in the different accessions of the olive collection subset are also important. They could have a profound impact on the aroma of the oils despite the fact that they are present at low concentrations in the oils (average 100 ng/g oil, range of 20–740 ng/g oil). The 2 and 3-methyl-butanal contents suggest that these BC aldehydes contribute decisively to the VOO aroma, since they have OAV values higher than one in all the accessions. Their aroma significance for VOO is especially notable because they are located in the ripe fruit sector of the SSW [26]. However, 2-methyl-butan-1-ol, which is responsible for the fusty off-flavor of the oil [27], was found below its odor threshold in all the cultivars assessed (**Table 2**). Thus, it should not contribute to the aroma of the oils. Finally, limonene seems not to be an important contributor to VOO aroma since only three cultivars of the olive collection subset had an OAV above one for this terpene (**Table 2**).

Factor analysis allowed explaining the pattern of correlations within different volatile compounds in the olive oils. **Figure 3** displays the factor analysis considering only those factors with eigenvalues higher than one and using the normalized Varimax method. Those factors explained 77.46% of the total variance. First factor explained 21.73% of the variance and second factor 12.60%. As shown, most of the C6 compounds spread along Factor 2 (green ellipse), while most of the C5 compounds distributed along Factor 1. Three different groupings may be distinguished for the latter (**Figure 2**, pink ellipses), which correspond to C5/LnA alcohols, aldehydes and pentene dimers, respectively. The group of C5/LnA alcohols partially overlaps the space of the C6 compounds. This indicates a greater metabolic proximity than other C5/LnA subgroups, which would support the hypothesis that C5/LnA alcohols are the first products formed from the homolytic branch of the LOX pathway [42], parallel to the pentene dimers formation, and finally oxidized enzymatically to C5/LnA carbonyls. Moreover, it was found previously that the pentene dimer content did not correlate with the C5/LA carbonyl or alcohol contents, suggesting that pentene dimers were only produced from LnA [44]. Accordingly, those compounds are located in **Figure 3** far from the group of pentene dimers.

On the other hand, the esters are grouped together overlapping the space of the C6 compounds, where the main precursors of esters are located [(hexan-1-ol (6C-9) and (Z)-hex-3-enol (6C-7)]. This might be indicative of a disconnection with the mainstream LOX pathway, which reflects the limitation of alcohol synthesis during VOO production due to the inactivation of ADH during the oil extraction process as we have previously demonstrated [45].

The PCA approach explains the correlations among the different classes of volatile compounds assessed in the oils of the olive collection subset and allows us to distinguish those olive cultivars that are especially rich in some of them (**Figure 4**). The first two PCs carried

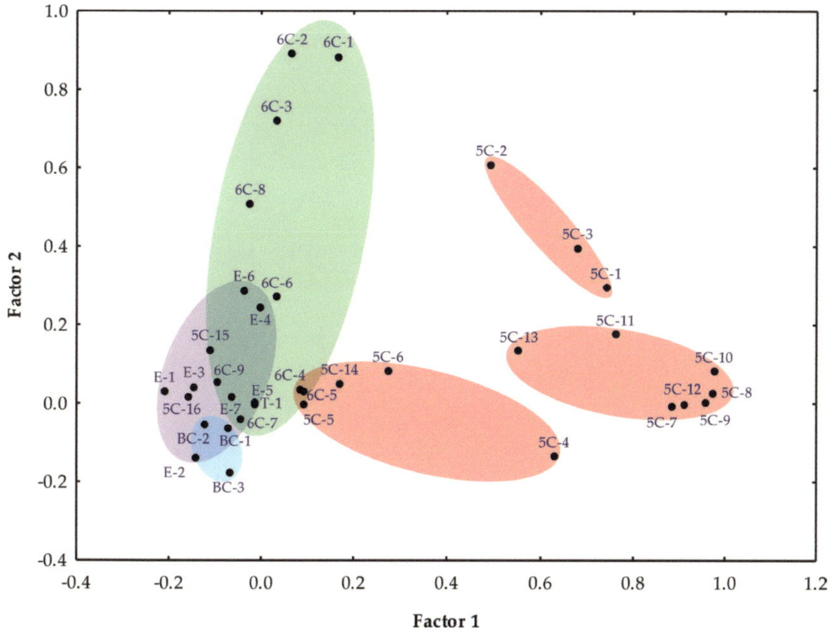

Figure 3. Factor analysis. Position of the main volatile compounds in the oils from the WOGC cultivar subset on the first two factors using the normalized Varimax method.

a moderate amount of important information, with the first factor and second factor correspondingly explaining 21.46% and 15.14% of the variance. We previously found quite similar values when assessing the content of the main volatile compounds in the oils of the progeny of the cross of cultivars Picual and Arbequina (21.03% and 17.29% for factor 1 and 2, respectively) [33]. Cultivars giving rise to oils with high C6/LnA aldehyde contents are situated around the center of the plot, while those producing oils with high C5/LnA contents are located along the negative axis of the second factor. Moreover, the oils from cultivars located along the first factor axis would have high C6/LnA aldehydes content as well as high levels of LOX esters or BC aldehydes. These compounds might act synergistically to increase the green-fruity odor notes. Most of the compounds included in those volatile classes are located in the green and ripe fruit sectors of the SSW [26]. Thus, it is possible to select cultivars from the germplasm collection subset, whose oils have a potential dominant green aroma, such as those cultivars located in the bisector of the first quadrant.

Table 2 shows that only a few volatile compounds are present in the olive cultivar subset at levels indicating that they may contribute to the oil aroma (OAV > 1). However, other volatile compounds contribute to just a given number of the oil cultivars. PCA was performed considering as variables only those volatile compounds that might contribute to the aroma of the oil in more than 10% of the cultivars (**Figure 5**). Most of these compounds were considered desirable for the aroma of VOO, except for hexan-1-ol (6C-9) and pent-1-en-3-one (5C-1), which provide unpleasant sensations according to literature [10, 26, 27]. The volatile compounds used as variables in **Figure 5** have been previously identified as key contributors to the aroma of various monovarietal olive oils from Spain, Italy, Argentina or Iran [46–49]. Factors 1 and 2 explain 46% of the data variation, quite similar to what was found for the progeny of the cross

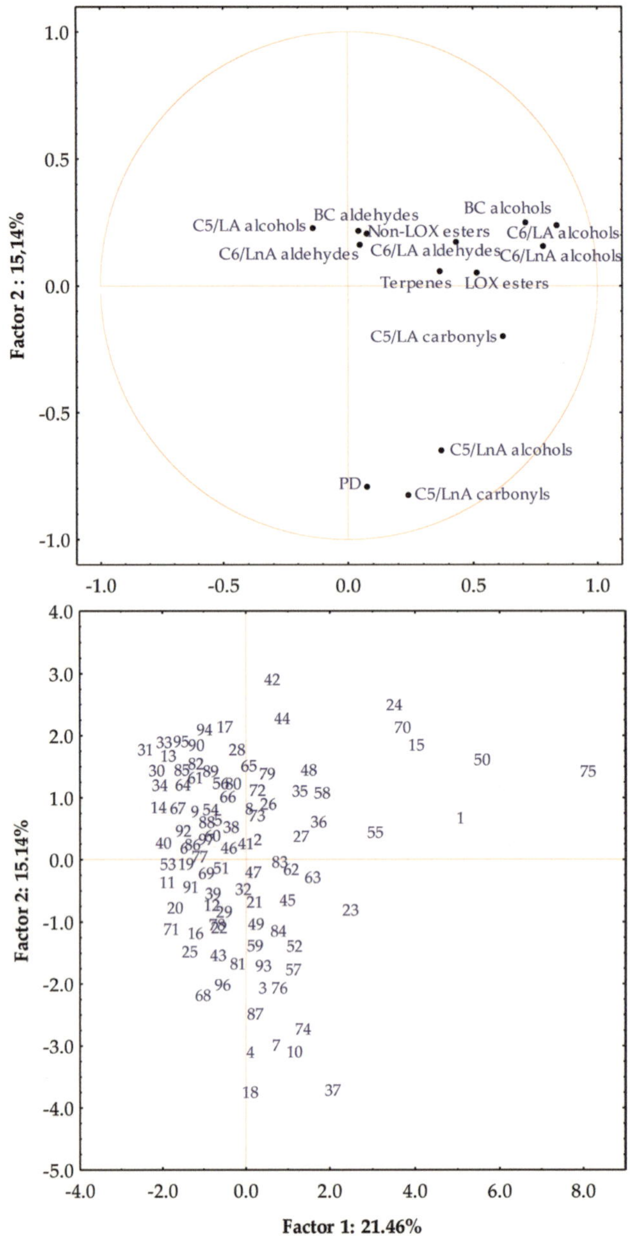

Figure 4. Principal component analysis of the main groups of volatile compounds in the oils from the WOGC cultivar subset. Up, vector distribution of the volatile compounds. Bottom, distribution of the cultivars. References for the cultivars are listed in **Table 1**.

of cultivars Picual and Arbequina [33]. When using as variables the volatile compounds that contribute to the aroma (OAV > 1) of the oils, the cultivars are distributed mainly along the bisectors of the second and fourth quadrants. The distribution of vectors in **Figure 5** allows identifying cultivars that presumably give rise to oils with remarkable sensory properties.

http://dx.doi.org/10.5772/intechopen.76343

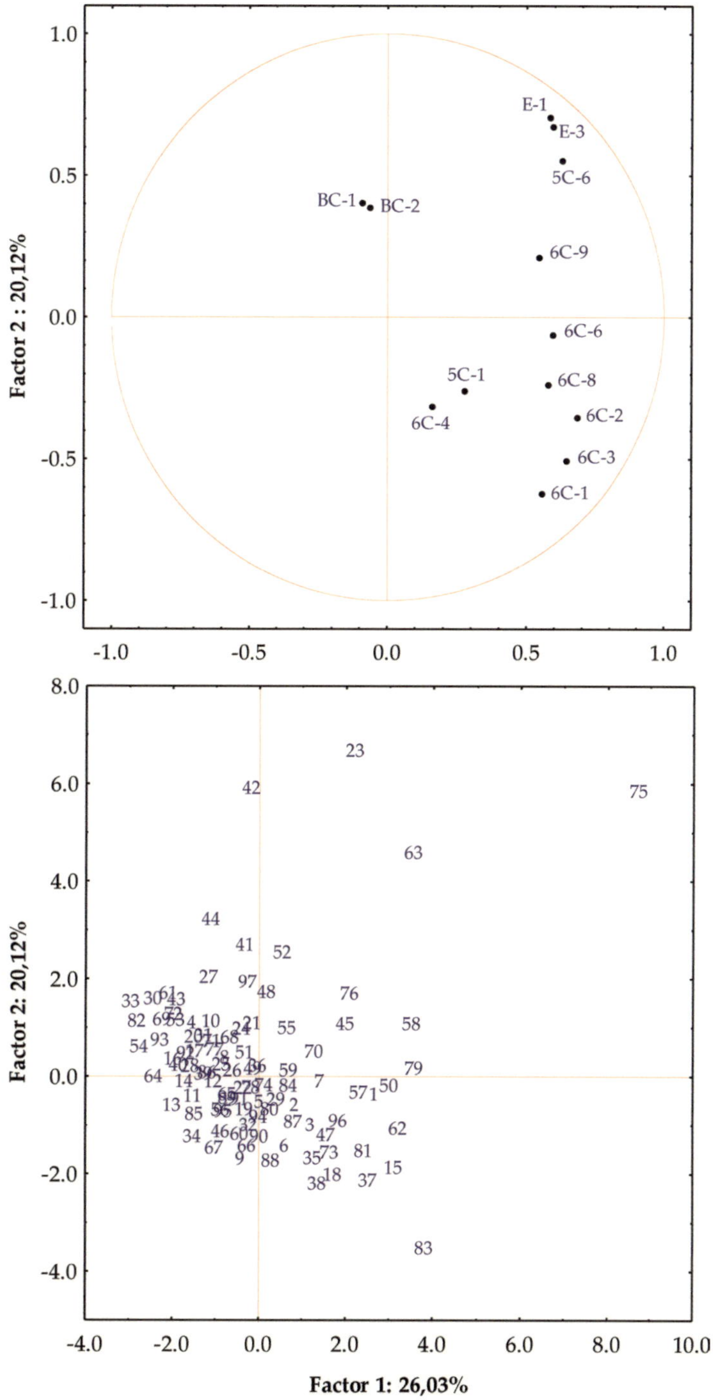

Figure 5. Principal component analysis of the volatile compounds contributing to the aroma (OAV > 1) of the WOGC cultivar subset. Up, vector distribution of the volatile compounds. Codes for the volatile compounds are listed in **Table 2**. Bottom, distribution of the cultivars from the collection. References for the cultivars are listed in **Table 1**.

4. Conclusions

Data demonstrated that the olive species presents a high level of variability in terms of the volatile fraction of the oils and, presumably, of the aroma quality. This aroma variability and the high genetic diversity of the cultivar germplasm collection suggest that it is possible both to identify old olive cultivars that give rise to oils with a high organoleptic quality and to select optimal parents for olive breeding programs with the aim of finding new cultivars with improved oil aroma. Multivariate analysis seems to be a particularly interesting tool for this purpose.

Acknowledgements

Funding for this research came from the OLEAGEN project of Genoma España and the project AGL2011-24442 from the Programa Nacional de Recursos y Tecnologías Agroalimentarias, both financed by the Spanish Government. The plant materials evaluated here were obtained from the cooperative breeding program carried out at the University of Cordoba, and at the Institute of Agricultural and Fishery Research and Training (IFAPA), Spain.

Conflict of interest

The authors declare no conflicts of interest.

Author details

Carlos Sanz[1*], Angjelina Belaj[2], Mar Pascual[1] and Ana G. Pérez[1]

*Address all correspondence to: carlos.sanz@ig.csic.es

1 Department of Biochemistry and Molecular Biology of Plant Products, Instituto de la Grasa, CSIC, Seville, Spain

2 IFAPA, Centro Alameda del Obispo, Cordoba, Spain

References

[1] Lucas L, Russell A, Keast R. Molecular mechanisms of inflammation. Anti-inflammatory benefits of virgin olive oil and the phenolic compound oleocanthal. Current Pharmaceutical Design. 2011;**17**:754-768

[2] Estruch R, Ros E, Salas-Salvadó J, Covas MI, Corella D, et al. Primary prevention of cardiovascular disease with a Mediterranean diet. New England Journal of Medicine. 2013;**368**:1279-1290

[3] Visioli F, Bernardini E. Extra virgin olive oil's polyphenols: Biological activities. Current Pharmaceutical Design. 2011;**17**:786-804

[4] Konstantinidou V, Covas MI, Muñoz-Aguayo D, Khymenets O, de La Torre R, et al. In vivo nutrigenomic effects of VOO polyphenols within the frame of the Mediterranean diet: A randomized trial. FASEB Journal. 2010; **24**:2546-2557

[5] Pérez-Jiménez F, Ruano J, Pérez-Martínez P, López-Segura F, López-Miranda J. The influence of olive oil on human health: Not a question of fat alone. Molecular Nutrition & Food Research. 2007;**51**:1199-1208

[6] Psaltopoulou T, Kosti RI, Haidopoulos D, Dimopoulos M, Panagiotakos DB. Olive oil intake is inversely related to cancer prevalence: A systematic review and a meta-analysis of 13800 patients and 23340 controls in 19 observational studies. Lipids in Health and Disease. 2011;**10**:127

[7] Genovese A, Caporaso N, Villani V, Paduano A, Sacci R. Olive oil phenolic compounds affect the release of aroma compounds. Food Chemistry. 2015;**181**:284-294

[8] Olías JM, Pérez AG, Rios JJ, Sanz C. Aroma of virgin olive oil: Biogenesis of the green odour notes. Journal of Agricultural and Food Chemistry. 1993;**41**:2368-2373

[9] Morales MT, Aparicio R, Ríos JJ. Dynamic headspace gas chromatographic method for determining volatiles in virgin olive oil. Journal of Chromatography A. 1994;**68**:455-462

[10] Angerosa F, Mostallino R, Basti C, Vito R. Virgin olive oil odour notes: Their relationships with the volatile compound from the lipoxigenase pathway and secoiridoid compounds. Food Chemistry. 2000;**68**:283-287

[11] Salas J, Williams M, Harwood JL, Sánchez SJ. Lipoxygenase activity in olive (Olea europaea) fruit. Journal of the American Oil Chemists Society. 1999;**76**:1163-1169

[12] Salas JJ, Sánchez J. Hydroperoxide lyase from olive (Olea europaea) fruits. Plant Science. 1999;**143**:19-26

[13] Salas JJ, Sánchez J. Alcohol dehydrogenases from olive (Olea europaea) fruit. Phytochemistry. 1998;**48**:35-40

[14] Salas JJ. Characterization of alcohol acyltransferase from olive fruit. Journal of Agricultural and Food Chemistry. 2004;**52**:3155-3158

[15] Gardner HW, Grove MJ, Salch YP. Enzymic pathway to ethyl vinyl ketone and 2-pentenal in soybean preparations. Journal of Agricultural and Food Chemistry. 1996;**44**:882-886

[16] Fisher AJ, Grimes HD, Fall R. The biochemical origin of pentenol emissions from wounded leaves. Phytochemistry. 2003;**62**:159-163

[17] Sánchez-Ortiz A, Pérez AG, Sanz C. Cultivar differences on nonesterified polyunsaturated fatty acid as a limiting factor for biogenesis of virgin olive oil aroma. Journal of Agricultural and Food Chemistry. 2007;**55**:7869-7873

[18] Sánchez-Ortiz A, Romero-Segura C, Sanz C, Pérez AG. Synthesis of volatile compounds of virgin olive oil is limited by the lipoxygenase activity load during the oil extraction process. Journal of Agricultural and Food Chemistry. 2012;**60**:812-822

[19] Belaj A, Dominguez-García MC, Atienza SG, Martín-Urdíroz N, de la Rosa R, Satovic Z, et al. Developing a core collection of olive (Olea europaea L) based on molecular markers (DArTs, SSRs, SNPs) and agronomic traits. Tree Genetics & Genomes. 2012;**8**:365-378

[20] León L, Beltrán G, Aguilera MP, Rallo L, Barranco D, de la Rosa R. Oil composition of advanced selections from an olive breeding program. Journal of Lipid Science and Technology. 2011;**113**:870-875

[21] Rjiba I, Gazzah N, Dabbou S, Hammami M. Evaluation of virgin olive oil minor compounds in progenies of controlled crosses. Journal of Food Biochemistry. 2011;**35**:1413-1423

[22] Lavee S. Aims, methods and advances in breeding of new olive (Olea europaea L.) cultivars. Acta Horticulturae. 1989;**286**:23-40

[23] Morales MT, Aparicio-Ruiz R, Aparicio R. Chromatographic methodologies: Compounds for olive oil odour issues. In: Aparicio R, Harwood J, editors. Handbook of Olive Oil: Analysis and Properties. New-York: Springer; 2013. p. 261

[24] The BPZ. Seven primary hexenols and their olfactory characteristics. Journal of Agricultural and Food Chemistry. 1971;**19**:1111-1114

[25] Hatanaka A, Kajiwara T, Horino H, Inokuchi KI. Odour-structure relationships in n-hexanols and n-hexenales. Zeitschrift fur Naturforschung Section A-A. 1992;**47**:183-189

[26] Aparicio R, Morales MT. Sensory wheels: A statistical technique for comparing QDA panels. Application to virgin olive oil. Journal of the Science of Food and Agriculture. 1995;**67**:247-257

[27] Angerosa F, Lanza B, Marsilio V. Biogenesis of "fusty" defect in virgin olive oils. Grasas y Aceites. 1996;**47**:142-150

[28] Aparicio R, Morales MT, Alonso MV. Relationship between volatile compounds and sensory attributes of olive oils by sensory wheel. Journal of the American Oil Chemists Society. 1996;**73**:1253-1264

[29] Commission Regulation 640/2008/EC. Amending Regulation (EEC) No 2568/91 on the characteristics of olive oil and olive-residue oil and on the relevant methods of analysis. Official Journal of the European Union. 2008;**L178**:11-16

[30] Angerosa F. Influence of volatile compounds on virgin olive oil quality evaluated by analytical approaches and sensor panels. European Journal of Lipid Science and Technology. 2002;**104**:639-660

[31] Morales MT, Luna G, Aparicio R. Comparative study of virgin olive oil sensory defects. Food Chemistry. 2005;**91**:293-301

[32] El Riachy M, Priego-Capote F, Rallo L, Luque de Castro MD, León L. Phenolic profile of virgin olive oil from advanced breeding selections. Spanish Journal of Agricultural Research. 2012;**10**:443-453

[33] Perez AG, de la Rosa R, Pascual M, Sanchez-Ortiz A, Romero-Segura C, Leon L, Sanz C. Assessment of volatile compound profiles and the deduced sensory significance of virgin olive oils from the progeny of Picual x Arbequina cultivars. Journal of Chromatography A. 2016;**1428**:305-315

[34] Cortés-Delgado A, Sánchez AH, de Castro A, López-López A, Montaño A. Volatile profile of Spanish-style green table olives prepared from different cultivars grown at different locations. Food Research International. 2016;**83**:131-142

[35] López-López A, Sánchez AH, Cortés-Delgado A, de Castro A, Montaño A. Relating sensory analysis with SPME-GC-MS data for Spanish-style green table olive aroma profiling. LWT-Food Science and Technology. 2018;**89**:725-734

[36] Cimato A, Dello Monaco D, Monaco CDC, Epifani M, Sani G. Analysis of single-cultivar extra virgin olive oils by means of an electronic nose and HS-SPME/GC/MS methods. Sensors and Actuators B: Chemical. 2006;**114**:674-680

[37] Melucci D, Bendini A, Tesini F, Barbieri S, Toschi TG. Rapid direct analysis to discriminate geographic origin of extra virgin olive oils by flash gas chromatography electronic nose and chemometrics. Food Chemistry. 2016;**204**:263-273

[38] Fortini M, Migliorini M, Cherubini C, Cecchi L, Calamai L. Multiple internal standard normalization for improving HS-SPME-GC-MS quantitation in virgin olive oil volatile organic compounds (VOO-VOCs) profile. Talanta. 2017;**165**:641-652

[39] Luna G, Morales MT, Aparicio R. Characterisation of 39 varietal virgin olive oils by their volatile composition. Food Chemistry. 2006;**98**:243-252

[40] Reiners J, Grosch W. Odourants of virgin olive oils with different flavor profiles. Journal of Agricultural and Food Chemistry. 1998;**46**:2754-2763

[41] Peres F, Jelen HH, Majcher MM, Arraias M, Ferreira-Dias S. Characterization of aroma compounds in Portuguese extra virgin olive oils from Galega vulgar and Cobrançosa cultivars using GC–O and GC×GC–ToFMS. Food Research International. 2007;**54**:1979-1986

[42] Angerosa F, Camera L, d'Alessandro N, Mellerio G. Characterization of seven new hydrocarbon compounds present in the aroma of virgin olive oils. Journal of Agricultural and Food Chemistry. 1998;**46**:648-653

[43] Kesen S, Kelebek H, Sen K, Ulas M, Selli S. GC–MS–olfactometric characterization of the key aroma compounds in Turkish olive oils by application of the aroma extract dilution analysis. Food Research International. 2013;**54**:1987-1994

[44] García-Vico L, Belaj A, Sanchez-Ortiz A, Martínez-Rivas JM, Pérez AG, Sanz C. Volatile compound profiling by HS-SPME/GC-MS-FID of a core olive cultivar collection as a tool for aroma improvement of virgin olive oil. Molecules. 2017;**22**:141

[45] Sánchez-Ortiz A, Romero-Segura C, Gazda VE, Graham IA, Sanz C, Pérez AG. Factors limiting the synthesis of virgin olive oil volatile esters. Journal of Agricultural and Food Chemistry. 2012;**60**:1300-1307

[46] Cecchi T, Alfei B. Volatile profiles of Italian monovarietal extra virgin olive oils via HS-SPME–GC–MS: Newly identified compounds, flavors molecular markers, and terpenic profile. Food Chemistry. 2013;**141**:2025-2035

[47] Reboredo-Rodríguez P, González-Barreiro C, Cancho-Grande B, Simal-Gándara J. Dynamic headspace/GC–MS to control the aroma fingerprint of extra-virgin olive oil from the same and different olive varieties. Food Control. 2012;**25**:684-695

[48] Fernandez MA, Assof M, Jofre V, Silva MF. Volatile profile characterization of extra virgin olive oils from Argentina by HS-SPME/GC-MS and multivariate pattern recognition tools. Food Analytical Methods. 2014;**7**:2122-2136

[49] Amanpour A, Kelebek H, Kesen S, Selli S. Characterization of aroma-active compounds in iIranian cv. mMari olive oil by aroma extract dilution analysis and GC–MS-Oolfactometry. Journal of the American Oil Chemists′ Society. 2016;**93**:1595-1603

Chapter 5

High-Order Calibration and Data Analysis in Chromatography

Hai-Long Wu, Xiao-Dong Sun, Huan Fang and
Ru-Qing Yu

Additional information is available at the end of the chapter

http://dx.doi.org/10.5772/intechopen.78624

Abstract

Multiway data analysis and tensorial calibration are gaining widespread acceptance with the rapid development of multichannel chromatographic instruments. By combining chromatographic techniques with chemometrics based on high-order calibration methods, some traditional problems in analysis, such as complicated pretreatment steps, long elution times, or even worse analysis results, can be avoided/improved. This chapter presents an overview from second-order to third-order data that cover theories and applications together with corresponding data processing in chromatography.

Keywords: chemometrics, second-order advantages, trilinear model, high-performance liquid chromatography-diode array detection, liquid chromatography-mass spectrometry

1. Introduction

Chromatography, first employed in Russia by the Italian-born scientist Mikhail Tsvet in 1900, is a laboratory technique for the separation of a mixture. In the first decade of the twentieth century, scientists continued to work with chromatography primarily for the purpose of separating plant pigments such as chlorophyll, carotenes, and xanthophylls. Since these pigments separated showed various colors (green, orange, and yellow, respectively), they gave the technique its name. After various types of chromatography had sprung up in the 1930s and 1940s, it became useful for many separation processes. Up to now, many chromatographic techniques have been developed, and they can be classified according to different properties. Based on chromatographic bed shape techniques, they can be divided into column chromatography and planar chromatography. Also, gas chromatography and liquid chromatography are classified by the physical state of mobile phase. In addition, there are also many other categories classified by other properties (i.e. separation mechanism, special techniques); but the chromatographic classification is out of scope of this chapter.

The purpose of chromatography is to separate the components of a mixture for later use. The mixture is dissolved in a fluid called the mobile phase, which carries it through a structure holding another material called the stationary phase. The separation is based on differential partitioning between the mobile and stationary phases. Subtle differences in a compound's partition coefficient result in differential retention on the stationary phase and thus affect the separation.

Nowadays, due to its prominent separation properties, chromatography techniques have become an indispensable tool for the routine analysis and research in pharmaceutical, biomedical, food, and environmental industries [1]. However, there are two main drawbacks needed to be solved/improved. The first one is about the sample itself; when complex matrix samples are analyzed, some proper tedious pretreatment procedures, such as extraction and purification, are necessary to remove the potential interferences contained in complex matrices. Optimizing these procedures is rather tedious and large sum of solvents' consumption are inevitable, making this method become uneconomical and environmentally unfriendly. What's more, in traditional chromatography analysis, when a complex sample is analyzed, the overlap between the analytes and matrix constituents is frequently observed; consequently, a long time or much more complex chromatography condition is required for the separation. In general, the elution time for each sample often costs 30–50 min, which is quite time consuming and inefficient. In the same time, some other problems such as baseline drift, changes in the shape of the peaks, incomplete extraction of the analytes, and shifts in the elution times may also decrease the quality of the final result of the analysis. Another problem with chromatography is due to its universal aspect. There are now hundreds of different chromatographic columns, which can be obtained from the market and new ones are being developed constantly [2, 3]. However, when faced with the large number of possible columns, it is hard for analysts to select which could be the most appropriate one for a given condition. Meanwhile, many laboratories and public institutions may not possess all available stationary phases and the column performance may become worse during long-term storage and/or usage in LC analysis [4]. Thus, the analysts often waste a lot of time in search of the most appropriate one from several different stationary phases for analysis. All of the above shortcomes may hinder the further development for chromatographic applications.

A current trend in quantitative analysis is to avoid tedious sample preprocessing steps and long chromatographic elution, exploiting the ability of modern data processing tools for mathematical resolution of coeluting components [5]. The combination of suitable chemometric tools along with chromatographic-spectral data or chromatographic-mass data may solve/improve the problem. With less time and solvent consumption, better quantitative results can be obtained. The multi-way (second- and third-order) calibration based on "mathematical separation" is a dazzling pearl in the field of chemical analysis and can calibrate the potential interferences and resolve coeluting peaks successfully in real samples with minimum sample preparation steps. Accurately concentration profiles of individual components of interest can also be obtained. This property generally refers to the prominent "second-order advantage," which has enormous potential in multiway analysis and becomes a recent focus of theoretical research and practical uses. Combining chromatography with multiway calibration has some distinct advantages because it can simplify the tedious multistep pretreatment and exploration of complicated chromatography separate conditions, showing the potential abilities for the analysis of various samples with different interferences at a time. Tedious pretreatment or purification procedures can be discarded by using prominent "mathematic separation" instead of tradition "physical/chemical separation." HPLC coupled with second-order calibration methods is especially popular for it can rapidly and simultaneously determine multiple compounds in complex backgrounds with unknown interferences, resolve coeluted peaks, and remove baselines drifts [6–10].

So far, a lot of algorithms for decomposition of multiway data arrays have already been proposed and genuinely provided alternative tools to analytical chemists for the convenient study of the body of multiway data arrays. Several methodologies have also been expounded in "Encyclopedia of Analytical Chemistry" [11] and "Factor Analysis in Chemistry" [12] at some length. To help readers systematically and intensively understand about concerning algorithms, a detailed description including multilinear models, the multiway cyclic symmetry property, the algorithms for multiway calibration, the estimation of the chemical rank, the toolbox for multiway calibration, and other fundamental issues and applications in chromatography has been presented in this paper.

2. Terminology and nomenclature in multiway data

To facilitate understanding for readers when dealing with multivariate analysis on multiway data arrays, it is necessary to introduce the terminology and nomenclature used in multiway data in the following.

2.1. Terminology

The relationship and difference between the concepts of "data order" and "data way" should be investigated firstly. The term "order" is the dimensions for data of a single sample and term "way" represents the data arrays stacked by all samples with similar properties. As shown in **Figure 1**, zeroth order corresponds to instruments producing a single response per sample, such as the reading of a pH meter or the absorbance at a single wavelength. First-order data are arranged as a vector or first-order tensor for a single sample, such as UV, fluorescence, infrared, and nuclear magnetic resonance spectra. At the same time, second-order data are formed when matrix data can be obtained for a single sample. There are two ways that second-order data can be obtained: (i) using a single instrument such as excitation-emission spectrofluorimeter (EEMs) or diode-array spectrophotometer to monitor the kinetics of a chemical reaction and (ii) using the hyphenated instruments such as high performance liquid chromatography-photodiode array detection (HPLC-DAD) or liquid chromatography-mass spectrometry (LC-MS). When the second-order data that obtained from a series of samples

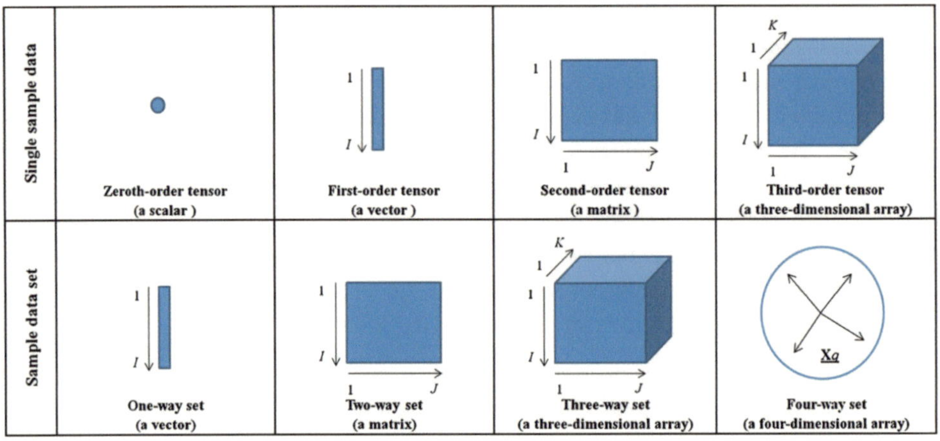

Figure 1. Relationships and differences between the concepts of "data order" and "data way" described with symbols.

(calibration and prediction samples) are stacked in one direction, three-dimensional array, which is also called as three-way array can be obtained, and the corresponding data are usually known as three-way data. Hence, when a series of samples are stacked into a single, zeroth-order (a scalar), first-order (a vector), second-order (a matrix), third-order (a three-way array), and higher order tensors can yield the corresponding one-way, two-way, three-way, four-way, and *N*-way data sets, respectively. The zeroth-order tensor calibration is also called as univariate calibration. This method has great restraint on its application as it needs full selectivity for the signals of target analytes. Except univariate calibration for the analysis of data, others are known as multivariate calibration, the analysis of second-order tensor and higher order tensor is denoted as multiway multivariate or multicomponent calibration.

Meanwhile, a detailed description for various sample types is also provided. Based on different functions, samples can be divided into calibration, prediction, and actual sets. Actual sets include predicted samples (the target analyte(s) is (are) unambiguously included) and/or real samples (whether the target analyte(s) is (are) included or unknown). Constituents present in the samples used for calibration and validation are regularly called "known" or "expected," which is expected in these sets as they are expected to be existed in actual samples. The expected constituents can be further divided into "calibrated" and "uncalibrated." The concentrations of former ones used for calibration are predesigned and known, while those of "calibrated" components in actual sets can also be available, involving the analyte(s) of interest. On the other hand, the constituents which are only included in actual sets are called "unknown" or "unexpected" and also potential interferences.

2.2. Nomenclature

In this chapter, lowercase italics represent scalars; two-way matrices are denoted by bold capitals; underlined bold capitals designate three-way arrays, the superscript T represents the transpose of a matrix, and the superscript + is the Moore-Penrose generalized inverse of a matrix. $||\cdot||_F$ designates the Frobenius matrix norm. To have a better understanding about

http://dx.doi.org/10.5772/intechopen.78624

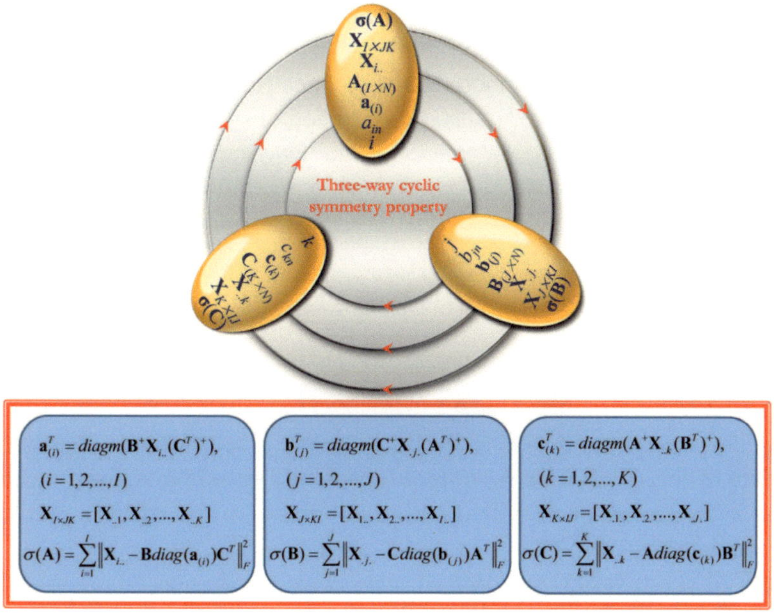

Figure 2. Schematic representation for three-way cyclic symmetry property.

the multiway calibration, readers are advised to comprehend an inner cyclic symmetry property of trilinear decomposition proposed by our laboratory in 1996 and also called as three-way cycle symmetry. As shown in **Figure 2**, elements, vectors, subscripts, and physical modes in resolved matrices, sliced matrices, and unfolded matrices, together with residue and resolution formulas, all obey the principle of inner cyclic symmetry property, circumrotating along the same way. **Table 1** provides the detailed information of the nomenclature mentioned.

X	**Three-way data array**
I, J, K	The three dimensions of three modes of **X**
x_{ijk}	The ijkth element of **X**
$\mathbf{A}_{I \times N}, \mathbf{B}_{J \times N}, \mathbf{C}_{K \times N}$	The three underlying profile matrices of **X with** $I \times N$, $J \times N$, and $K \times N$, respectively
a_{in}, b_{jn}, c_{kn}	The inth, jnth, and knth elements of the three underlying profile matrices **A**, **B**, and **C**, respectively
$\mathbf{a}_{(i)}, \mathbf{b}_{(j)}, \mathbf{c}_{(k)}$	The ith, jth, and kth row vectors of profile matrices **A**, **B**, and **C**, respectively
diag($\mathbf{a}_{(i)}$), diag ($\mathbf{b}_{(j)}$), diag($\mathbf{c}_{(k)}$)	Diagonal matrices with elements equal to the elements of $\mathbf{a}_{(i)}$, $\mathbf{b}_{(j)}$, and $\mathbf{c}_{(k)}$, respectively
$\mathbf{X}_{i..}, \mathbf{X}_{.j.}, \mathbf{X}_{..k}$	The ith horizontal, jth lateral, and kth frontal slices of **X**, respectively
$\mathbf{E}_{i..}, \mathbf{E}_{.j.}, \mathbf{E}_{..k}$	The ith horizontal, jth lateral, and kth frontal slices of the three-way array residue **E**, respectively
$\mathbf{e}_{ijk}$	The ijkth element of the three-way residue array **E**

Table 1. Detailed information of the nomenclature mentioned.

Similar to the three-way cyclic symmetry, the four-way and five-way cyclic symmetries for quadrilinear and quinquelinear decomposition can be obtained easily by simple mathematical manipulation of exchanging the symbols similar to three-way cyclic symmetry. These regularities provide useful instructions for the standardization of symbol systems in multiway data analysis, for better understanding the essence of multiway multilinear decomposition, developing new multiway calibration algorithms, and exploring multilinear algebra in mathematics.

3. Theory

3.1. Multilinear models

According to the data type and its inner cyclic symmetry property, the multilinear models can be divided into trilinear, quadrilinear, quinquelinear, and even higher linear models. In chromatographic analysis combined with multiway calibration, the trilinear and quadrilinear models are commonly used.

3.1.1. Trilinear model

Harshman [13] together with Carroll and Chang [14] first proposed the PARAFAC (PARallel FACtor analysis) model with the name of CANDECOMP in the year of 1970. In this trilinear model, each element x_{ijk} of a three-way array **X** ($I \times J \times K$) can be reasonably fit to the following equation:

$$x_{ijk} = \sum_{n=1}^{N} a_{in} b_{jn} c_{kn} + e_{ijk}, i = 1, 2, \ldots, I; j = 1, 2, \ldots, J; k = 1, 2, \ldots, K. \quad (1)$$

where N represents the total number of detectable components including the component(s) of interest, uncalibrated background(s), and unknown interferences. **Figure 3** illustrates the graphical representation of a trilinear model of three-way data array **X**. **A**, **B**, and **C** are the three underlying profile matrices of X with $I \times N$, $J \times N$, and $K \times N$, respectively; **I** is the three-way diagonal core array of size $N \times N \times N$ with ones on the superdiagonal and zeros

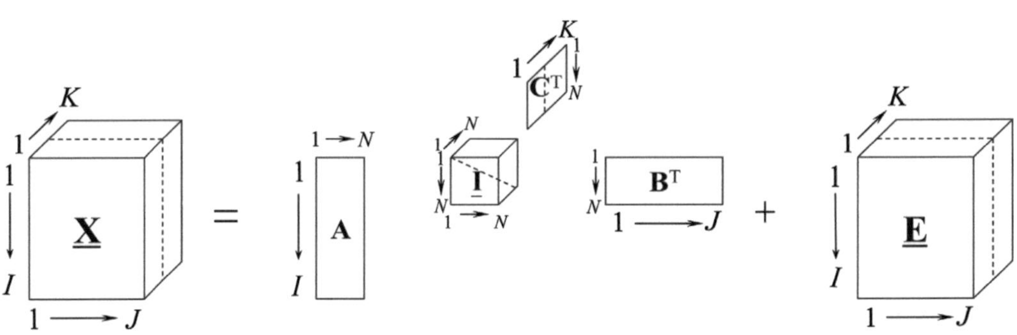

Figure 3. Schematic representation of trilinear model.

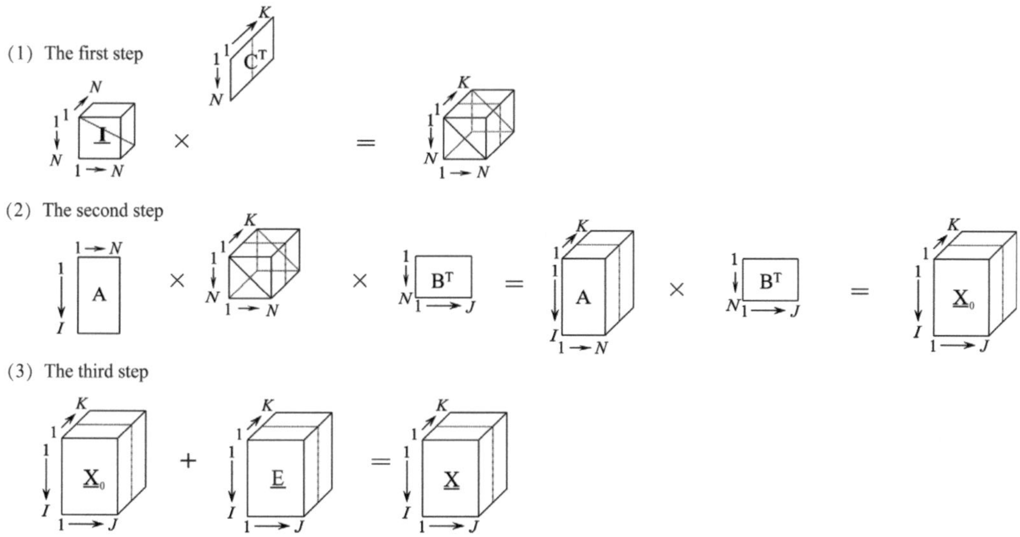

Figure 4. The inverse procedures to return three-way data array X.

elsewhere; and ***E*** is the three-way residue data array of size $I \times J \times K$. To further comprehend the mathematic meaning of the trilinear model graphically expressed, the response data array **X** is returned with three inverse steps as described in **Figure 4**.

3.1.2. Quadrilinear model

Considering a model of the real-valued four-way data array **X** ($I \times J \times K \times L$), in which each element x_{ijk} can be expressed as [8, 15]:

$$x_{ijkl} = \sum_{n=1}^{N} a_{in} b_{jn} c_{kn} d_{ln} + e_{ijkl}, \quad i = 1, 2, \ldots, I; j = 1, 2, \ldots, J; k = 1, 2, \ldots, K; l = 1, 2, \ldots, L. \quad (2)$$

where a_{in}, b_{jn}, c_{kn}, and d_{ln} correspond to the underlying profile matrices $\mathbf{A}_{I \times N}$, $\mathbf{B}_{J \times N}$, $\mathbf{C}_{K \times N}$, and $\mathbf{D}_{L \times N}$ of **X** ($I \times J \times K \times L$), respectively. The term e_{ijkl} is the element of the four-way residual array **E** ($I \times J \times K \times L$). Then, the modeled part of x_{ijkl} is quadrilinear in the parameter sets a_{in}, b_{jn}, c_{kn}, and d_{ln}. The graphical representation of a quadrilinear model of four-way data array **X** is shown in **Figure 5**.

3.2. Data preprocessing

The correctness of decomposition of a multilinear model requires that the multilinear model holds multilinearity. However, there are some nonmultilinear factors which can cause a multilinear model to deviate the multilinearity. For example, in the chromatography type of trilinear model such as HPLC-DAD and LC-MS data, the time shift and baseline problem among different runs will cause the trilinear model to deviate the trilinearity. Thus, the data

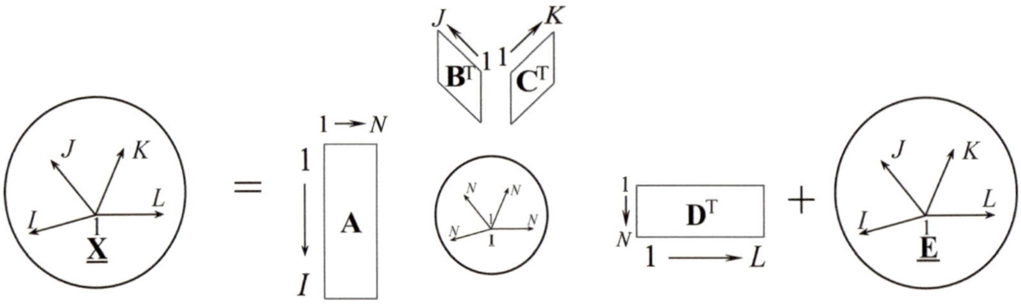

Figure 5. Schematic representation of quadrilinear model.

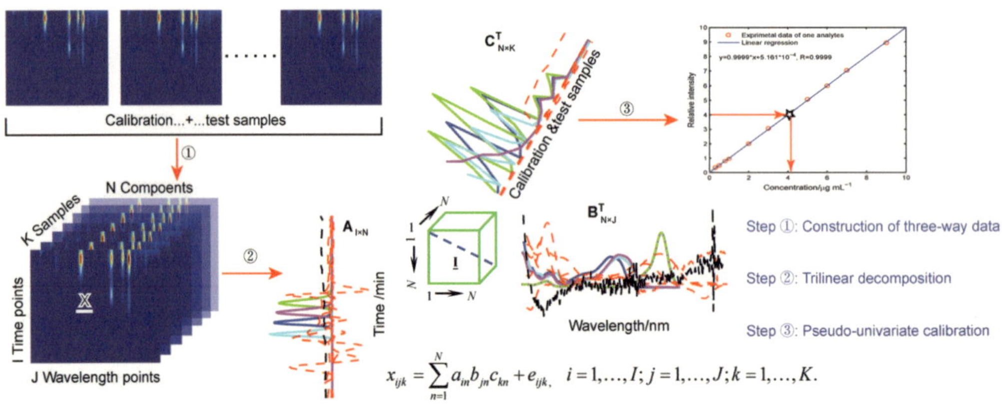

Figure 6. Schematic representation of entire chemometrics-assisted LC-DAD analytical strategy.

arrays in multivariate calibration must need appropriate data preprocessing procedures before a multilinear decomposition. The schematic representation of entire chemometrics-assisted LC-DAD and LC-MS analytical strategy is shown in **Figures 6** and **7**, respectively.

3.3. Algorithm

3.3.1. ATLD

The ATLD algorithm is a universal second-order calibration method for decomposition of three-way data arrays. It is based on an alternating least squares principle without any constrains and an improved iterative procedure that utilizes the Moore-Penrose generalized inverse based on singular value decomposition. It has been widely used in three-way data analysis due to the advantages of being insensitive to excessive component numbers and fast convergence.

According to its cyclic symmetry property, the trilinear model can also be expressed in matrix notation as follows:

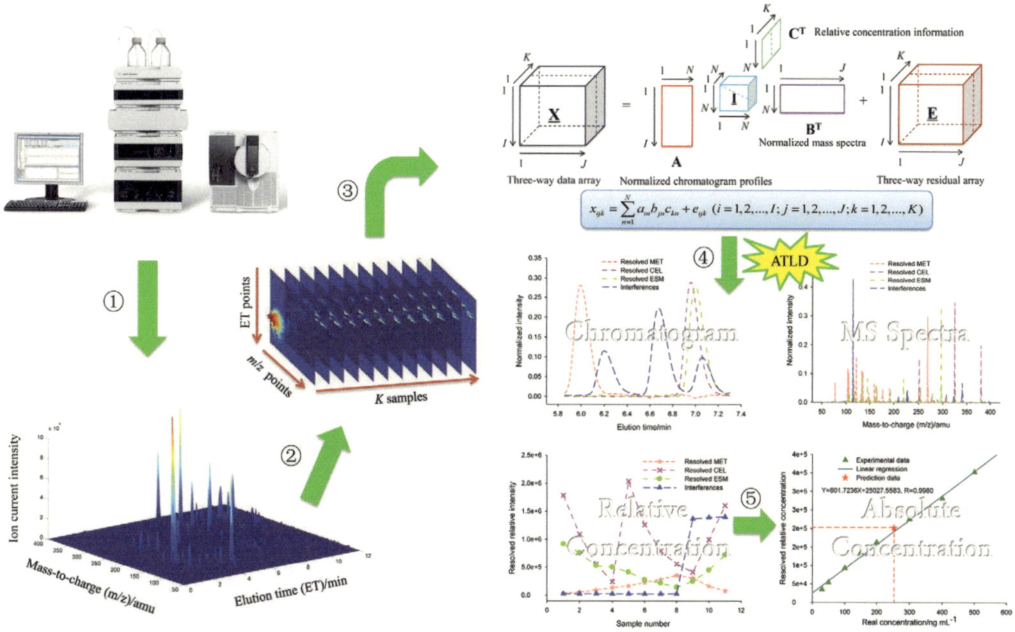

Figure 7. Schematic representation of entire chemometrics-assisted LC-MS analytical strategy.

$$\mathbf{X}_{i..} = \mathbf{B}diag\left(\mathbf{a}_{(i)}\right)\mathbf{C}^{\mathrm{T}} + \mathbf{E}_{i..}, \text{ for } i = 1, 2, \dots, I, \tag{3}$$

$$\mathbf{X}_{.j.} = \mathbf{C}diag\left(\mathbf{b}_{(j)}\right)\mathbf{A}^{\mathrm{T}} + \mathbf{E}_{.j.}, \text{ for } j = 1, 2, \dots, J, \tag{4}$$

$$\mathbf{X}_{..k} = \mathbf{A}diag\left(\mathbf{c}_{(k)}\right)\mathbf{B}^{\mathrm{T}} + \mathbf{E}_{..k}, \text{ for } k = 1, 2, \dots, K, \tag{5}$$

Due to the property called as cyclic symmetry of the trilinear model, the three expressions are equal to each other in mathematics. According to Eqs. (3)–(5), the loss function to be minimized is the sum of the squares of the elements of the residual matrices, which can be expressed as:

$$\sigma\left(\mathbf{a}_{(i)}\right) = \sum_{i=1}^{I} \left\|\mathbf{X}_{i..} - \mathbf{B}diag\left(\mathbf{a}_{(i)}\right)\mathbf{C}^{\mathrm{T}}\right\|_{\mathrm{F}}^{2}, \tag{6}$$

$$\sigma\left(\mathbf{b}_{(j)}\right) = \sum_{j=1}^{J} \left\|\mathbf{X}_{.j.} - \mathbf{C}diag\left(\mathbf{b}_{(j)}\right)\mathbf{A}^{\mathrm{T}}\right\|_{\mathrm{F}}^{2}, \tag{7}$$

$$\sigma\left(\mathbf{c}_{(k)}\right) = \sum_{k=1}^{K} \left\|\mathbf{X}_{..k} - \mathbf{A}diag\left(\mathbf{c}_{(k)}\right)\mathbf{B}^{\mathrm{T}}\right\|_{\mathrm{F}}^{2}, \tag{8}$$

By using the loss functions abovementioned, ATLD alternately minimizes the three objective functions over **C** on fixed **A** and **B**, over **A** on fixed **B** and **C**, and then over **B** on fixed **C** and **A**.

The updates for the three profile matrices (A, B, and C) are based on the least squares principle and can be represented as follows:

$$\mathbf{a}_{(i)}^{\mathrm{T}} = diagm\left(\mathbf{B}^{+}\mathbf{X}_{i..}\left(\mathbf{C}^{\mathrm{T}}\right)^{+}\right), \text{ for } i = 1, 2, \ldots, I, \tag{9}$$

$$\mathbf{b}_{(j)}^{\mathrm{T}} = diagm\left(\mathbf{C}^{+}\mathbf{X}_{.j.}\left(\mathbf{A}^{\mathrm{T}}\right)^{+}\right), \text{ for } j = 1, 2, \ldots, J, \tag{10}$$

$$\mathbf{c}_{(k)}^{\mathrm{T}} = diagm\left(\mathbf{A}^{+}\mathbf{X}_{..k}\left(\mathbf{B}^{\mathrm{T}}\right)^{+}\right), \text{ for } k = 1, 2, \ldots, K, \tag{11}$$

herein $diagm(\cdot)$ stands for a column N-vector and its elements are diagonal elements in square matrix. In every iteration cycle, **A** and **B** are normalized column-wise with unit length. With the help of the resolved profile matrices **C**, we can get the concentrations of analytes of interests in actual samples via regression of the appropriate column of **C** corresponding to each analyte against its standard concentrations.

Due to the operation based on sliced matrices with less size and two other major strategies, ATLD holds the fastest convergence. The truncated least squares method employs the tolerance to truncate the small singular values in the singular value decomposition. In addition, selecting diagonal elements makes ATLD retain trilinearity property indeed and be insensitive to the excessive estimation of component numbers. The advantages have been reviewed by Fleming and Kowalski [16]. Based on the above advantages, it is very suitable to handle second-order data obtained from hyphenated instruments, such as HPLC-DAD, LC-MS, and GC-MS.

3.3.2. SWATLD

The SWATLD algorithm, as a derivative of ATLD, is also widely employed as it can yield better results in many cases. It alternately minimizes three objective functions with intrinsic relationship and also holds the characteristics of fast convergence and being insensitive to excessive component numbers. The detail explanations of these properties have been provided by authors in the original paper [17]. Three new residues can be expressed as:

$$\mathbf{B}^{+}\mathbf{X}_{i..} = diag\left(\mathbf{a}_{(i)}\right)\mathbf{C}^{\mathrm{T}} + \mathbf{B}^{+}\mathbf{E}_{i..}, \quad \mathbf{X}_{i..}\left(\mathbf{C}^{+}\right)^{\mathrm{T}} = \mathbf{B}diag\left(\mathbf{a}_{(i)}\right) + \mathbf{E}_{i..}\left(\mathbf{C}^{+}\right)^{\mathrm{T}}, \text{ for } i = 1, 2, \ldots, I, \tag{12}$$

$$\mathbf{C}^{+}\mathbf{X}_{.j.} = diag\left(\mathbf{b}_{(j)}\right)\mathbf{A}^{\mathrm{T}} + \mathbf{C}^{+}\mathbf{E}_{.j.} \quad \mathbf{X}_{.j.}\left(\mathbf{A}^{+}\right)^{\mathrm{T}} = \mathbf{C}diag\left(\mathbf{b}_{(j)}\right) + \mathbf{E}_{.j.}\left(\mathbf{A}^{+}\right)^{\mathrm{T}}, \text{ for } j = 1, 2, \ldots, J, \tag{13}$$

$$\mathbf{A}^{+}\mathbf{X}_{..k} = diag\left(\mathbf{c}_{(k)}\right)\mathbf{B}^{\mathrm{T}} + \mathbf{A}^{+}\mathbf{E}_{..k}, \quad \mathbf{X}_{..k}\left(\mathbf{B}^{+}\right)^{\mathrm{T}} = \mathbf{A}diag\left(\mathbf{c}_{(k)}\right) + \mathbf{E}_{..k}\left(\mathbf{B}^{+}\right)^{\mathrm{T}}, \text{ for } k = 1, 2, \ldots, K, \tag{14}$$

By introducing some reasonable weight terms, three new objective functions are established and can be expressed as follows:

$$\begin{aligned} \mathrm{S}(\mathbf{A}) = &\sum_{i=1}^{I}\left\|\left(\mathbf{B}^{+}\mathbf{X}_{i..} - diag\left(\mathbf{a}_{(i)}\right)\mathbf{C}^{\mathrm{T}}\right)^{\mathrm{T}} \times diag\left(sqrt\left(\mathbf{1}./diagm\left(\mathbf{C}^{\mathrm{T}}\mathbf{C}\right)\right)\right)\right\|_{\mathrm{F}}^{2} \\ &+ \sum_{i=1}^{I}\left\|\left(\mathbf{X}_{i..}\left(\mathbf{C}^{\mathrm{T}}\right)^{+} - \mathbf{B}diag\left(\mathbf{a}_{(i)}\right)\right) \times diag\left(sqrt\left(\mathbf{1}./diagm\left(\mathbf{B}^{\mathrm{T}}\mathbf{B}\right)\right)\right)\right\|_{\mathrm{F}}^{2}, \end{aligned} \tag{15}$$

$$\begin{aligned} \mathrm{S}(\mathbf{B}) &= \sum_{j=1}^{J} \left\| \left(\mathbf{C}^{+}\mathbf{X}_{.j.} - diag(\mathbf{b}_{(j)})\mathbf{A}^{\mathrm{T}}\right)^{\mathrm{T}} \times diag\left(sqrt\left(\mathbf{1}./diagm\left(\mathbf{A}^{\mathrm{T}}\mathbf{A}\right)\right)\right) \right\|_{\mathrm{F}}^{2} \\ &+ \sum_{j=1}^{J} \left\| \left(\mathbf{X}_{.j.}\left(\mathbf{A}^{\mathrm{T}}\right)^{+} - \mathbf{C}diag(\mathbf{b}_{(j)})\right) \times diag\left(sqrt\left(\mathbf{1}./diagm\left(\mathbf{C}^{\mathrm{T}}\mathbf{C}\right)\right)\right) \right\|_{\mathrm{F}}^{2}, \end{aligned} \tag{16}$$

$$\begin{aligned} \mathrm{S}(\mathbf{C}) &= \sum_{k=1}^{K} \left\| \left(\mathbf{A}^{+}\mathbf{X}_{..k} - diag(\mathbf{c}_{(k)})\mathbf{B}^{\mathrm{T}}\right)^{\mathrm{T}} \times diag\left(sqrt\left(\mathbf{1}./diagm\left(\mathbf{B}^{\mathrm{T}}\mathbf{B}\right)\right)\right) \right\|_{\mathrm{F}}^{2} \\ &+ \sum_{k=1}^{K} \left\| \left(\mathbf{X}_{..k}\left(\mathbf{B}^{+}\right)^{\mathrm{T}} - \mathbf{A}diag(\mathbf{c}_{(k)})\right) \times diag\left(sqrt\left(\mathbf{1}./diagm\left(\mathbf{A}^{\mathrm{T}}\mathbf{A}\right)\right)\right) \right\|_{\mathrm{F}}^{2}. \end{aligned} \tag{17}$$

Due to the unique optimizing strategy, this algorithm is more efficient than others. It can provide more satisfactory results than ATLD with moderate noise levels. Moreover, it can deal with the problem of moderate collinearity, but it is not so effective when data are collinear severely.

3.3.3. APTLD

The APTLD algorithm was developed by Xia et al. [18], and it can provide some improved properties. It alternately minimizes three new least squares-based objective functions by using the constraint functions as penalty terms of the PARAFAC error. Eqs. (12)–(14) are the new objective functions, which alternately used as the constraint terms. By introducing large penalty terms and combining them with residue functions (18)–(20) to establish three objective functions, APTLD transforms these constrained problems into non-constrained ones. Then, it alternately minimizes the following three objective functions to resolve the model:

$$\begin{aligned} \mathrm{S}(\mathbf{A}) &= \sum_{k=1}^{K} \left\| \mathbf{X}_{..k} - \mathbf{A}diag(\mathbf{c}_{(k)})\mathbf{B}^{\mathrm{T}} \right\|_{\mathrm{F}}^{2} + q\Big(\sum_{j=1}^{J} \left\| diag\left(sqrt\left(\mathbf{1}./diagm\left(\mathbf{B}^{\mathrm{T}}\mathbf{B}\right)\right)\right)\left(\mathbf{C}^{+}\mathbf{X}_{.j.} - diag(\mathbf{b}_{(j)})\mathbf{A}^{\mathrm{T}}\right) \right\|_{\mathrm{F}}^{2} \\ &+ \sum_{k=1}^{K} \left\| \left(\mathbf{X}_{..k}\left(\mathbf{B}^{\mathrm{T}}\right)^{+} - \mathbf{A}diag(\mathbf{c}_{(k)})\right) diag\left(sqrt\left(\mathbf{1}./diagm\left(\mathbf{C}^{\mathrm{T}}\mathbf{C}\right)\right)\right) \right\|_{\mathrm{F}}^{2}\Big), \end{aligned} \tag{18}$$

$$\begin{aligned} \mathrm{S}(\mathbf{B}) &= \sum_{i=1}^{I} \left\| \mathbf{X}_{i..} - \mathbf{B}diag(\mathbf{a}_{(i)})\mathbf{C}^{\mathrm{T}} \right\|_{\mathrm{F}}^{2} + r\Big(\sum_{k=1}^{K} \left\| diag\left(sqrt\left(\mathbf{1}./diagm\left(\mathbf{C}^{\mathrm{T}}\mathbf{C}\right)\right)\right)\left(\mathbf{A}^{+}\mathbf{X}_{..k} - diag(\mathbf{c}_{(k)})\mathbf{B}^{\mathrm{T}}\right) \right\|_{\mathrm{F}}^{2} \\ &+ \sum_{i=1}^{I} \left\| \left(\mathbf{X}_{i..}\left(\mathbf{C}^{\mathrm{T}}\right)^{+} - \mathbf{B}diag(\mathbf{a}_{(i)})\right) diag\left(sqrt\left(\mathbf{1}./diagm\left(\mathbf{A}^{\mathrm{T}}\mathbf{A}\right)\right)\right) \right\|_{\mathrm{F}}^{2}\Big), \end{aligned} \tag{19}$$

$$\begin{aligned} \mathrm{S}(\mathbf{C}) &= \sum_{j=1}^{J} \left\| \mathbf{X}_{.j.} - \mathbf{C}diag(\mathbf{b}_{(j)})\mathbf{A}^{\mathrm{T}} \right\|_{\mathrm{F}}^{2} + p\Big(\sum_{i=1}^{I} \left\| \mathrm{diag}\left(sqrt\left(\mathbf{1}./diagm\left(\mathbf{A}^{\mathrm{T}}\mathbf{A}\right)\right)\right)\left(\mathbf{B}^{+}\mathbf{X}_{i..} - diag(\mathbf{a}_{(i)})\mathbf{C}^{\mathrm{T}}\right) \right\|_{\mathrm{F}}^{2} \\ &+ \sum_{j=1}^{J} \left\| \left(\mathbf{X}_{.j.}\left(\mathbf{A}^{\mathrm{T}}\right)^{+} - \mathbf{C}diag(\mathbf{b}_{(j)})\right) diag\left(sqrt\left(\mathbf{1}./diagm\left(\mathbf{B}^{\mathrm{T}}\mathbf{B}\right)\right)\right) \right\|_{\mathrm{F}}^{2}\Big), \end{aligned} \tag{20}$$

where p, q, and r represent penalty factors. The performance of APTLD depends on the choice of the penalty factor values. When the values are very small, it will lead to a lot of iterations and sensitivity to excess factors, which is close to that of PARAFAC algorithm; particularly, when $p = q = r = 0$, APTLD can be regarded as a variant of PARAFAC. However, this algorithm will become insensitive to excess factors and speed up convergence when larger values of p, q, and r are selected. According to the variance among different trials and computational burdens, a further increase in p, q, and r values will make APTLD perform theoretically better. Therefore, its performance can be exquisitely improved by adjusting the penalty factors p, q, and r on the basis of particular circumstances and special needs.

3.3.4. APQLD

The APQLD algorithm [19] as an extension of APTLD for decomposition of quadrilinear data is applied to third-order calibration. Similar to APTLD, four objective functions can be obtained as:

$$\begin{aligned}\mathrm{S}(\mathbf{A}) &= \sum_{k=1}^{K}\sum_{l=1}^{L}\|\mathbf{X}_{..kl} - \mathbf{A}diag(\mathbf{d}_{(l)})diag(\mathbf{c}_{(k)})\mathbf{B}^{\mathrm{T}}\|_{\mathrm{F}}^{2} \\ &+ q\Big(\sum_{k=1}^{K}\sum_{l=1}^{L}\Big\|\Big(\mathbf{X}_{..kl}(\mathbf{B}^{\mathrm{T}})^{+} - \mathbf{A}diag(\mathbf{d}_{(l)})diag(\mathbf{c}_{(k)})\Big)sqrt(\mathbf{W}_{\mathrm{B}})\Big\|_{\mathrm{F}}^{2} \\ &+ \sum_{j=1}^{J}\sum_{k=1}^{K}\|sqrt(\mathbf{W}_{\mathrm{D}})(\mathbf{D}^{+}\mathbf{X}_{jk..} - diag(\mathbf{c}_{(k)})diag(\mathbf{b}_{(j)})\mathbf{A}^{\mathrm{T}})\|_{\mathrm{F}}^{2}\Big),\end{aligned} \tag{21}$$

$$\begin{aligned}\mathrm{S}(\mathbf{B}) &= \sum_{l=1}^{L}\sum_{i=1}^{I}\|\mathbf{X}_{i..l} - \mathbf{B}diag(\mathbf{a}_{(i)})diag(\mathbf{d}_{(l)})\mathbf{C}^{\mathrm{T}}\|_{\mathrm{F}}^{2} \\ &+ r\Big(\sum_{l=1}^{L}\sum_{i=1}^{I}\Big\|\Big(\mathbf{X}_{i..l}(\mathbf{C}^{\mathrm{T}})^{+} - \mathbf{B}diag(\mathbf{a}_{(i)})diag(\mathbf{d}_{(l)})\Big)sqrt(\mathbf{W}_{\mathrm{C}})\Big\|_{\mathrm{F}}^{2} \\ &+ \sum_{k=1}^{K}\sum_{l=1}^{L}\|sqrt(\mathbf{W}_{\mathrm{A}})(\mathbf{A}^{+}\mathbf{X}_{..kl} - diag(\mathbf{d}_{(l)})diag(\mathbf{c}_{(k)})\mathbf{B}^{\mathrm{T}})\|_{\mathrm{F}}^{2}\Big),\end{aligned} \tag{22}$$

$$\begin{aligned}\mathrm{S}(\mathbf{D}) &= \sum_{j=1}^{J}\sum_{k=1}^{K}\|\mathbf{X}_{.jk.} - \mathbf{D}diag(\mathbf{c}_{(k)})diag(\mathbf{b}_{(j)})\mathbf{A}^{\mathrm{T}}\|_{\mathrm{F}}^{2} \\ &+ p\Big(\sum_{j=1}^{J}\sum_{k=1}^{K}\Big\|\Big(\mathbf{X}_{.jk.}(\mathbf{A}^{\mathrm{T}})^{+} - \mathbf{D}diag(\mathbf{c}_{(k)})diag(\mathbf{b}_{(j)})\Big)sqrt(\mathbf{W}_{\mathrm{A}})\Big\|_{\mathrm{F}}^{2} \\ &+ \sum_{i=1}^{I}\sum_{j=1}^{J}\|sqrt(\mathbf{W}_{\mathrm{C}})(\mathbf{C}^{+}\mathbf{X}_{ij..} - diag(\mathbf{b}_{(j)})diag(\mathbf{a}_{(i)})\mathbf{D}^{\mathrm{T}})\|_{\mathrm{F}}^{2}\Big),\end{aligned} \tag{23}$$

where $\mathbf{W}_{\mathrm{A}} = diag(\mathbf{1}./diagm(\mathbf{A}^{\mathrm{T}}\mathbf{A}))$, $\mathbf{W}_{\mathrm{B}} = diag(\mathbf{1}./diagm(\mathbf{B}^{\mathrm{T}}\mathbf{B}))$, $\mathbf{W}_{\mathrm{C}} = diag(\mathbf{1}./diagm(\mathbf{C}^{\mathrm{T}}\mathbf{C}))$, and $\mathbf{W}_{\mathrm{D}} = diag(\mathbf{1}./diagm(\mathbf{D}^{\mathrm{T}}\mathbf{D}))$. APQLD algorithm decomposes the quadrilinear model by alternatively minimizing the four objective functions abovementioned. The performance of APQLD also depends on the selection of the penalty factors p, q, r, and s. Obviously, it can be considered as a variant of the four-way PARAFAC when the four penalty factors equal to 0.

APQLD retains the second-order advantage possessed by second-order calibration and holds additional advantage. By introducing a new fourth mode, it can relieve the serious problem of collinearity, which cannot be solved by three-way algorithms.

3.4. Rank estimation

It is always an important and intractable problem to estimate chemical ranks (the number of factors or components) for the trilinear model before decomposing a three-way data array. Theoretically, it can be seemingly solved by selecting the appropriate algorithms, which are insensitive to the excessive component numbers (chemical ranks). Nevertheless, these algorithms also guarantee that the component number (chemical rank) chosen should be no fewer than the underlying one. As a matter of fact, when the component number selected is far more than the actual one, it may lead to a model fitting error and a large deviation for the predicted results. On the contrary, the performances of the algorithm on providing accurate solutions will be largely improved when the most appropriate factors are chosen in analytical system.

Based on this, a lot of methods have been developed for estimating the chemical ranks. In general, they can be roughly fallen into two main categories. The first one is on the basis of the trilinear model, which includes split-half analysis [20], Wu's maximum rank method [21], core consistency diagnostic (CORCONDIA) [22], ADD-ONE-UP [23], and self-weighted alternating trilinear decomposition and Monte Carlo simulation (SWATLD-MCS) [24]. The core of split-half analysis concerns a relatively complex splitting skill, and hence the result depends on splitting schemes greatly. CORCONDIA and ADD-ONE-UP are two of the most commonly used methods in determining the chemical ranks. However, they are quite time consuming sometimes. Furthermore, the severe collinearity data may also lead to a heavy computation burden and even get error results. Self-weighted alternating trilinear decomposition and Monte Carlo simulation (SWATLD-MCS) operate in two main steps. First of all, Monte Carlo simulation is applied to generated one pseudo three-way data array. Sorted mean relative concentration values can then be obtained by applying SWATLD to decompose the three-way data array created by MCS. By comparing the sorted mean relative concentration value, this method can determine the chemical rank. The other ones belong to nonmodel methods such as orthogonal projection approach (OPA) [25], two-mode subspace comparison (TMSC) [26], factor indicator function (IND) [27], subspace projection of pseudo high-way array (SPPH) [28], linear transform method incorporating Monte Carlo simulation (LTMC) [29], and region based on moving windows subspace projection technique (RMWSPT) [30]. Though all of the above methods can be applied to rank estimation, it is impossible to find one among them which can guarantee the correct results under all situations. Actually, more than one method is often utilized in analysis to ensure the accuracy of the analytical results [8, 15].

3.4.1. *Maximum rank method*

The maximum rank method was firstly proposed by Wu et al. [21] to estimate the chemical rank for ATLD and ATLD's variants, as the following form shows:

$$rank\,\underline{\mathbf{X}} = max\{rank(\mathbf{X}_{I\times JK}), rank(\mathbf{X}_{J\times KI}), rank(\mathbf{X}_{K\times IJ})\}, \tag{24}$$

In practice, the number of factors will also be determined as follows:

$$rank\,\underline{\mathbf{X}} = max\left\{rank\left(\sum_{i=1}^{I}\mathbf{X}_{i..}\right), rank\left(\sum_{j=1}^{J}\mathbf{X}_{.j.}\right), rank\left(\sum_{k=1}^{K}\mathbf{X}_{..k}\right)\right\}, \tag{25}$$

$$rank\,\underline{\mathbf{X}} = max\left\{rank\left(\mathbf{X}_{I\times JK}\mathbf{X}_{I\times JK}^{T}\right), rank\left(\mathbf{X}_{J\times KI}\mathbf{X}_{J\times KI}^{T}\right), rank\left(\mathbf{X}_{K\times IJ}\mathbf{X}_{K\times IJ}^{T}\right)\right\}, \tag{26}$$

where rank (.) denotes the numerical rank estimate of a matrix based on a singular value decomposition procedure with a default tolerance. This method is universal and suitable to be used in any instance and can get satisfactory results when estimating the chemical rank of the three-way data.

3.4.2. ADD-ONE-UP

ADD-ONE-UP was proposed by Chen et al. in [23] for determining the chemical rank. It operates by fitting two reconstructed three-way data arrays by PARAFAC with a gradually increasing component numbers and then determines the chemical rank by examining the residual sum of squares (SSR). The method is convenient and powerful, and some nonideal experimental conditions (such as slight collinearity and unknown backgrounds) can be handled.

1. Unfold the obtained three-way data array $\mathbf{X}$ into a two-way data set $\mathbf{X}_{I\times JK}$.
2. Decompose $\mathbf{X}_{I\times JK}$ by SVD, $\mathbf{X}_{I\times JK} = \mathbf{USV}^{T}$.
3. Define $\mathbf{X}_c = \mathbf{U}_c\mathbf{S}_c\mathbf{V}_c^{T}$, $\mathbf{U}_c$ and $\mathbf{V}_c$ consist of the first c columns of **U** and **V**, respectively; $\mathbf{S}_c$ is a diagonal matrix with diagonal elements equal to the first c diagonal elements of **S**.
4. Fold $\mathbf{X}_c$ into a three-way data array $\mathbf{X}_c$, then resolve it by PARAFAC with $N = c$ (c = 1, 2, 3, …,). The residual sum of squares is denoted by SSR_c.
5. Repeat steps 3 and 4 until SSR_c reaches its minimum or satisfies the equations below: $SSR_{c1} < s_{c1}^2$ and $SSR_{c1+1} > s_{c1+1}^2$ and $SSR_{c1+2} > s_{c1+2}^2$ (s_i represents the ith diagonal element of matrix **S** and s_{c1}^2 denotes the variance obtained by the inclusion of c1th component in the truncating step).
6. Unfold $\mathbf{X}$ in another dimension to obtain $\mathbf{X}_{IK\times J}$, then perform the same steps from 2 to 5 to get *c2*, which meets similar relationships like *c1*.
7. The factor numbers applied in decomposing the trilinear data array $\mathbf{X}$ should be the smaller one between *c1* and *c2*, i.e. F = min(*c1*, *c2*).

This method utilizes the eigenvalues of factor analysis and the residuals of trilinear decomposition. It can cope with nonideal experimental conditions like varying backgrounds and moderate collinearity. However, as it is based on the PARAFAC algorithm, ADD-ONE-UP has some drawbacks. It is rather time consuming due to the need to run PARAFAC for many times. Furthermore, this method may suffer from a heavy computational burden by reason of two-factor degeneracies and may yield inaccurate results.

3.4.3. CORCONDIA

The principle of CORCONDIA is to assess the similarity between the superdiagonal array $\mathbf{T}$ and the least squares-fitted $\mathbf{G}$ with a gradually increasing number of components. CORCONDIA is defined as:

$$core\ consistency = 100 \times \left(1 - \frac{\sum_{d=1}^{N}\sum_{e=1}^{N}\sum_{f=1}^{N}\left(g_{def} - t_{def}\right)^2}{\sum_{d=1}^{N}\sum_{e=1}^{N}\sum_{f=1}^{N} t_{def}^2}\right), \tag{27}$$

where g_{def} stands for the element of $\mathbf{G}$, t_{def} represents the elements of T, and N denotes the number of factors in the model.

For an ideal trilinear model, g_{def} is equal to t_{def} and the value of core consistency will be equal to 100%. Usually, the model can be regarded as "very trilinear" as the value of the core consistency above 90%, whereas a value nearly 50% will indicate a problematic model, which contains both trilinear and non-trilinear variations. A value close to zero or even negative means that the model is not valid. Although it is an effective method, it suffers from the drawbacks of PARAFAC.

3.5. Some related fundamental issues

3.5.1. Chromatographic peak alignment procedure

Chromatographic peak alignment is a challenge in the field of complex system analysis by multiway calibration methods. Some methods for peak alignment have been developed based on the second-order instruments, which generate a matrix data for per sample. These methods [31], for example iterative target factor analysis coupled to COW (ITTFA-COW), rank minimization (RM), parallel factor analysis alignment, and other recently proposed methods based on multivariate curve resolution-alternating least squares, employ signals of two-way structure to align chromatographic peaks shifts. In theory, these methods are aimed at the alignment of local chromatographic regions and therefore satisfactory results can be obtained for the time shifts existed in the whole chromatogram. They can achieve accurate time alignment regardless of the presence of unknown interferences. Not long ago, Yu and co-workers developed a new algorithm for chromatographic peak alignment, derived from the famous rank minimization method. It aligns time shift among samples and then utilizes trilinear decomposition algorithm to interpret the overlapping chromatographic peaks to quantify target analytes [31].

Figure 8(A) depicts the graphical representation of the rank minimization method (RM). A significant advantage of this method is that alignment can be successfully carried out even when the potential interferences coeluted with the analyte of interest. To have a better view on

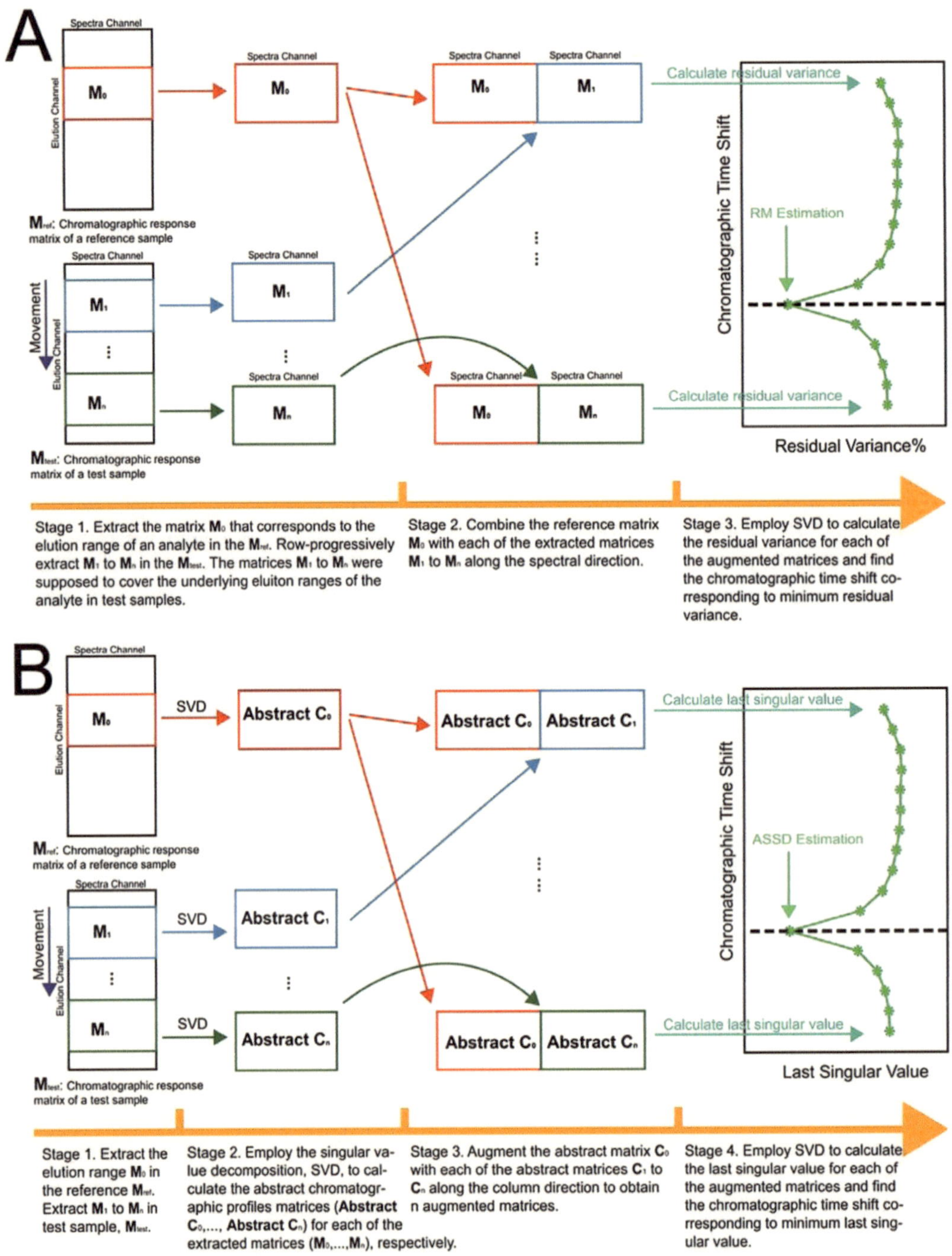

Figure 8. Graphical illustration of RM (A) and ASSD (B).

this method, a series of fixed-size time window (rectangles) along the retention time directions is applied in **Figure 8(A)**. In particular, the red rectangle $\mathbf{M}_0$ stands for the retention time range of analyte in the response of reference sample, and the retention time range between green and blue rectangles in the response of a test sample is the underlying time shift ranges of the analyte. By row-wisely moving the fixed-size time window on the test sample along the retention time direction, the rectangles from $\mathbf{M_1}$ to $\mathbf{M_n}$, can be extracted from the response of the test sample; then, augmented matrices, which are defined as $[\mathbf{M_0} \mid \mathbf{M_1}]$,…, $[\mathbf{M_0} \mid \mathbf{M_n}]$ [stage 2 in **Figure 8(A)**] in the retention time direction, can be obtained. Finally, the singular value decomposition is performed on these augmented matrices and results in a list of residual variance. Consequently, the percentage of the residual variance plotted against each chromatographic time shift will mark clearly the time shift point correction corresponding to the minimum residual variance.

The abstract subspace difference (ASSD) method uses abstract chromatographic profiles for alignment. Accordingly, the response matrix **X** can be expressed in the form of singular value decomposition (SVD) notations as follows:

$$\mathbf{X} = \mathbf{USV}^{\mathrm{T}} + \mathbf{E} \tag{28}$$

herein, the column vector **U** represents the abstract chromatographic profiles, while the V is the abstract spectra profiles; in the strict sense, all of them are not necessarily correspond to the real ones. Suppose that two data matrices have been collected: a reference data, $\mathbf{X}_{\mathrm{ref}}$, which includes only one analyte, and a test data, $\mathbf{X}_{\mathrm{test}}$, which collects the analyte together with other unknown interferences. Hence, based on the singular value decomposition, the abstract chromatographic profiles for reference and test samples can be acquired separately:

$$\mathbf{X}_{ref} = \mathbf{U}_{ref}\mathbf{S}_{ref}\mathbf{V}_{ref}{}^{\mathrm{T}} + \mathbf{E}_{ref}, \tag{29}$$

$$\mathbf{X}_{test} = \mathbf{U}_{test}\mathbf{S}_{test}\mathbf{V}_{test}{}^{\mathrm{T}} + \mathbf{E}_{test}. \tag{30}$$

In the ideal situation, no noise is present, and there is no time shift between reference and test samples. In this case, the mathematical rank of the augmented matrix $[\mathbf{U}_{ref} \mid \mathbf{U}_{test}]$ will be identical to that of $\mathbf{U}_{test}$. However, in the situations where the chromatographic retention time of the analyte is not the same for the reference and test samples, the mathematical rank of the augmented matrix, $[\mathbf{U}_{ref} \mid \mathbf{U}_{test}]$, will become larger than the actual ones. Therefore, the core of ASSD method is to look for the augmented matrix with minimum mathematical rank for alignment, which is the same as the rank minimization method, except that ASSD uses the abstract chromatographic profiles for alignment instead of the underlying ones.

Figure 8(B) shows the graphical illustration of the ASSD method. In order to calculate the abstract chromatographic profiles for each of the extracted matrices $\mathbf{M}_1$ to $\mathbf{M}_n$, an additional step, SVD, has been introduced in the Stage 1 of **Figure 8(B)**. Additionally, this new method uses the last singular value instead of the percentage of residual variance in the last stage to represent time shift correction. In practical measurement, aligning time shift for target analyte between the reference and a test sample according to the critical criterion of the mathematical rank of the augmented matrix is impractical. However, the augmented matrix, $[\mathbf{U}_{ref} \mid \mathbf{U}_{test}]$, will become a seriously ill-conditioned matrix provided that the time shift has been successfully aligned. Hereby, chromatographic peak alignment can be transformed to find the most ill-conditioned augmented matrix among the augmented matrices as shown in the Stage 3 of **Figure 8(B)**. As the total variance is the sum of the squared elements of the augment matrix, $[\mathbf{U}_{ref} \mid \mathbf{U}_{test}]$, it will be a steady state value and equal to the column numbers. Hence, a smaller last singular value will definitely correspond to a more ill-conditioned matrix.

3.5.2. Background drift

Non-trilinear factors such as background drift is unavoidable sometimes in the chromatographic analysis due to the composition of gradient elution and/or nature of complicated matrices, which may lead to wrong analysis results by the aforementioned chemometric algorithms. Amigo and co-workers have summarized the intuitive graphics and mathematical models used in handling chromatographic data issues [32]. Multivariate curve resolution (MCR) methods are typical examples.

A chromatographic background drift correction strategy [33] was developed in 2007 by our group for LC × LC × DAD data. The core idea is to perform trilinear decomposition, which is based on the alternating trilinear decomposition (ATLD) algorithm for the instrumental response data. In analysis, the background drift can be eliminated by regarding it as an extra component or factor. This method uses trilinear decomposition to resolve the raw data, to extract, and subtract the background component from the raw data for acquisition of the signal of analytes with a flat baseline. A detailed schematic description on how to subtract the background drift from raw three-way chromatographic data is illustrated in **Figure 9**.

Recently, a method that uses orthogonal spectral signal projection (OSSP) to simultaneously solve various kinds of chromatographic background drift was studied [33]. The analytical results indicated that OSSP coupled with PARAFAC can be used for handling coelution and background drift problems in chromatographic analysis. It indicates that more accurate analysis results can be obtained, regardless of the presence of background drift and unknown interferences.

4. Application

Based on the "second-order or high-order advantages" provided by chemometrics methods, some actual applications have been developed for the analysis of pharmaceuticals, biological matrices, foods, cosmetics, environmental matrices, and others. Multiway calibration algorithms have been employed to enhance the selectivity and can obtain accurate predicted concentration of analyte(s) of interest free from interference of potential interfering matrix. These applications summarized in **Table 2** are reviewed in the following six aspects.

4.1. Pharmaceuticals

In this field, two or three drugs have been simultaneously detected in aqueous solution or Chinese traditional medicine. The data analyzed are second-order tensors, which are obtained by high performance liquid chromatography-photodiode array detection (HPLC-DAD).

Su et al. proposed a method for simultaneously quantifying the main effective constituents such as puerarin, daidzin, and daidzein in traditional Chinese medicine kudzuvine root by using HPLC-DAD with ATLD algorithm [34].

Nowadays, traditional Chinese medicine (TCM) plays an important role in the healthcare system. Thus, considerable attention has been paid to Chinese patent medicine (CPM), which generally consists of several TCMs and other ingredients. It is significantly important to quantify the constituents of CPM and plasma for pharmacological analysis. Liu et al. determined two effective constituents, costunolide and dehydrocostuslactone, in plasma sample and Chinese patent medicine Xiang Sha Yang Wei capsule by using HPLC-DAD coupled with alternating trilinear decomposition (ATLD) algorithm [35].

Besides, Ding et al. determined isoniazid and pyrazinamide by using HPLC-DAD coupled with three different second-order calibration algorithms including ATLD, alternating fitting residue (AFR), and self-weighted alternating trilinear decomposition (SWATLD). The results showed that all the three algorithms could be used for solving overlapped chromatograms and unknown interferences successfully, and the analysis results obtained from AFR were slightly better in this situation [36].

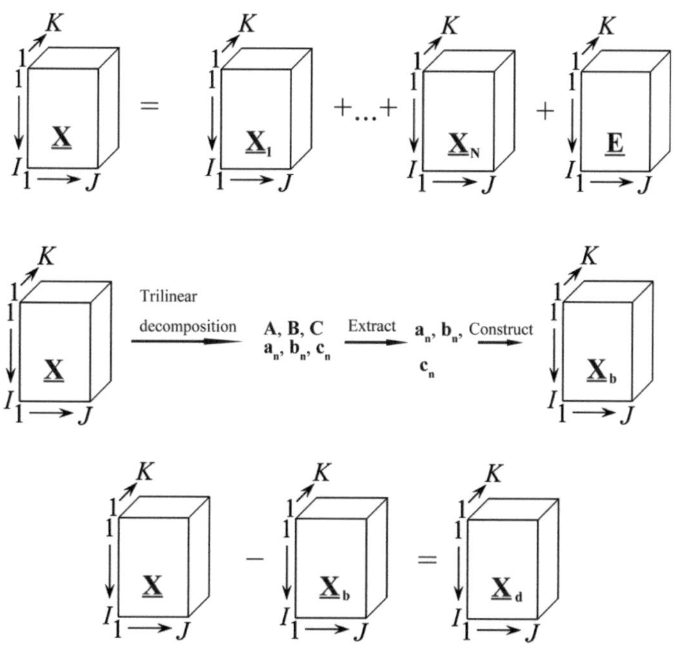

Figure 9. Schematic description on how to remove the background drift from three-dimensional instrumental data.

4.2. Biological matrices

Biological samples often contain various endogenous substances such as amino acids, hormones and neurotransmitters. Determining the concentrations of these molecules or metabolites is an integral part of clinical research and also helpful for understanding pathophysiology and mechanism of diseases. Human urine and plasma are commonly primary research systems.

High blood pressure, widely called hypertension, is a cardiac chronic disease with a symptom of sustaining rise in systemic arterial blood pressure. Zhao et al. carried out the simultaneous quantification of 11 antihypertensives, human serum, health product, and Chinese patent medicine samples by using HPLC-DAD with the aid of second-order calibration based on ATLD algorithm [37].

Tyrosine kinases are critical regulators of cell growth and differentiation growth and differentiation. The measurement of concentration of TKIs in different biofluids plays a significant role in optimizing the individual dosage regimen and reducing the risk of inapposite dosages. For the analysis of four tyrosine kinase inhibitors in different plasma samples, HPLC-DAD was utilized without absolutely chromatographic separations by resorting to ATLD algorithm. The contents of four tyrosine kinase inhibitors in different complex plasma samples can be accurately determined [38].

Liu et al. simultaneously determined vincristine, vinblastine, vindoline, catharanthine, and yohimbine in *Catharanthus roseus* and human serum samples utilizing ATLD algorithm to analyze the resulting three-way data array stacked by HPLC-DAD [39].

β-blockers are the first-line therapeutic agents for treating cardiovascular diseases and also a class of prohibited substances in athletic competitions. Therefore, rapid screening for multiple β-blockers in a single analysis has been of growing demand in clinical toxicology, forensic science, and antidoping control as well. Gu et al. proposed a smart strategy that combines three-way liquid chromatography-mass spectrometry (LC-MS) data with second-order calibration method based on alternating trilinear decomposition (ATLD) algorithm for simultaneous determination of 10 b-blockers in human urine and plasma samples [40]. The quantitative results were validated by the LC-MS/MS operated in multiple reaction monitoring (MRM) mode.

4.3. Foods

The applications in this field cover the analysis of contaminants, essential ingredients, and additives.

Synthetic phenolic antioxidants as food additives were successfully determined in edible vegetable oil by using HPLC-DAD and APTLD [9]. Some extraction procedures, in which the antioxidants of interest would be separated, is unnecessary and the 10 antioxidants can be eluted within 6 min.

Yin et al. proposed a smart strategy that combined HPLC-DAD with ATLD algorithm to solve varying interfering patterns from different chromatographic columns and sample matrices for the rapid simultaneous determination of six synthetic colorants in beverages with little sample pretreatment [1].

Type of data	Algorithm	Analytes	Ref.
Pharmaceuticals			
HPLC-DAD	ATLD	Puerarin, daidzin, and daidzein	[34]
HPLC-DAD	ATLD	Costunolide and dehydrocostuslactone	[35]
HPLC-DAD	ATLD, SWATLD, AFR	Isoniazid and pyrazinamide	[36]
Biological matrices			
HPLC-DAD	ATLD	Eleven antihypertensives	[37]
HPLC-DAD	ATLD	Four tyrosine kinase inhibitors	[38]
LC-MS	ATLD	Ten β-blockers	[40]
LC-MS	ATLD	Six antidiabetic agents	[48]
HPLC-DAD	ATLD	Five vinca alkaloids	[39]
Foods			
HPLC-DAD	PARAFAC, ATLD, SWATLD	Sudan I and Sudan II	[49]
HPLC-DAD	ATLD	Six synthetic colorants	[1]
HPLC-DAD	APTLD	Synthetic phenolic antioxidants	[9]
HPLC-DAD	ATLD,PCA	Eight coeluted compounds in tea	[7]
HPLC-DAD	ATLD	Eight flavonoids	[42]
HPLC-DAD	ATLD	nine polyphenols	[43]
HPLC-DAD	ATLD	Twelve quinolones	[44]
HPLC-DAD	ATLD, PCA-LDA	Thirteen phenolic compounds	[45]
HPLC-DAD	ATLD	Twelve polyphenols	[46]
LC-MS	ATLD	Ten mycotoxins	[47]
Environmental matrices			
HPLC-DAD	SWATLD	Three pre-emergence herbicides	[50]
HPLC	ATLD	1-Chloro-2,4-dinitrobenzene and 3,5-dinitrobenzoic acid	[51]
HPLC	ATLD	Five dimethylphenol isomers	[53]
HPLC	ATLD	Catechol, resorcinol and hydroquinone	[52]

Table 2. Reviewed applications.

Tea is one of the most widely consumed beverages in the world. The biological functions of tea have been reported in numerous studies, such as anti-inflammation, antiatherosclerotic, antioxidant, anticarcinoma, antiobesity, and antiviral properties. These beneficial effects are related to the presence of purine alkaloids and polyphenols in tea. An attractive chemometrics-enhanced HPLC-DAD strategy was proposed by Yin et al. for simultaneous and fast determination of eight coeluted compounds including gallic acid, caffeine, and six catechins in 10 kinds of Chinese teas by using second-order calibration method based on ATLD algorithm [41]. Subsequently, based on the quantitative results, principal component analysis (PCA) was used to conduct a cluster analysis for these Chinese teas.

Propolis is a naturally occurring resinous hive product gathered by worker honeybees from buds and barks of different plant species. Sun et al. developed a fast analytical strategy by combining HPLC-DAD with ATLD algorithm for simultaneous determination of eight flavonoids in propolis capsule samples [42].

Honey is a wholesome natural food product well known for its high nutrition. The antioxidant ability of a number of honeys has been determined and found to be significantly correlated to the contents of polyphenols, which can affect the quality of honeys and their products beneficial for improving overall health and preventing some diseases. By using second-order calibration for development of HPLC-DAD method, Zhang et al. quantified nine polyphenols in five kinds of honey samples successfully [43]. Quinolones, a kind of antibacterial, which is widely used in agriculture for its high antimicrobial activity, were also detected by HPLC-DAD with ATLD algorithm in honey samples [44].

Wine phenolic compounds, as secondary metabolites and functional components, determine the important sensorial characteristics of wines, such as mouth-feel, fragrance, and color. The combination of HPLC-DAD and second-order calibration method based on ATLD has been used for the determination of 13 phenolic compounds in red wines, and linear discriminant analysis (PCA-LDA) was applied for distinguishing wines aged for years [45]. Similarly, the same strategy was carried out by Wang et al. for simultaneously quantify 12 polyphenols in different kinds of apple peel and pulp samples [46].

Mycotoxins are a class of highly carcinogenic substances often naturally occurring in the moldy foods, especially cereals. Liu et al. proposed a smart strategy that combines three-way LC-MS data with second-order calibration method based on ATLD algorithm for direct, fast, and interference-free determination of multiclass regulated mycotoxins in complex cereal samples [47]. Ten mycotoxins with different property could be fast eluted out and detected by full scanning MS with a segmented fragment program to enhance the sensitivity.

By using LC-MS in combination with second-order calibration method based on ATLD algorithm, Gu et al. simultaneously green determined six coeluted sulfonylurea-type oral antidiabetic agents in healthy herbal teas and human plasma samples [48]. The strategy proved to be a promising method for resolution and determination of coeluted multianalytes of interest in complex samples while avoiding elaborate sample pretreatment steps and complicated experimental conditions as well as more sophisticated high-cost instrumentations.

For the determination of Sudan dyes in hot chilli samples, HPLC-DAD was employed without completely chromatographic separations by using PARAFAC, ATLD, and SWATLD [49]. The low contents of Sudan I and Sudan II could be accurately determined in complex chilli mixtures.

4.4. Environmental matrices

In this field, we analyzed the analytes in aqueous solution, soil, tap water, river, and effluent water, mainly containing organic contaminants and pesticides.

Herbicides, which are chemicals often employed to kill weeds without causing injury to desirable vegetation, have been widely used. These may lead to their accumulation in the environment and cause continuous and serious pollution or even toxicity to crops and humans. Qing et al. developed a novel strategy for analysis of three pre-emergence herbicides in environment samples using HPLC-DAD with SWATLD algorithm [50].

Chemometrics-assisted HPLC-DAD strategy has a great potential in analysis of target analytes in complex environmental matrices. So far, this strategy has been utilized for determination of 1-chloro-2,4-dinitrobenzene and 3,5-dinitrobenzoic acid [51], catechol, resorcinol, and hydroquinone [52] as well as five dimethylphenol isomers [53] in environment successfully.

5. Conclusion

This chapter scientifically describes in detail the various multiway chemometrics methodologies and applications in chromatography. We have built more canonical symbol systems, noted the inner mathematical cyclic symmetry property for multilinear decomposition, introduced several multiway calibration algorithms, explored the rank estimation of multiway data array, and analyzed numerous actual systems by homemade methods. Some fundamental issues related to chromatographic analysis such as peak alignment and background drift were also discussed and solved. By combining chromatographic techniques with chemometrics based on multiway calibration methods, complicated and tedious sample pretreatment can be greatly simplified and long chromatographic elution can be avoided. All the applications abovementioned are universal, rapid, and sensitive for the determination of a variety of analytes in complex matrices.

Acknowledgements

The authors gratefully acknowledge the National Nature Science Foundation of China (Grant Nos. 21575039 and 21775039) and the Foundation for Innovative Research Groups of NSFC (Grant No. 21521063) for financial supports.

Conflict of interest

There are no conflicts to declare.

Author details

Hai-Long Wu*, Xiao-Dong Sun, Huan Fang and Ru-Qing Yu

*Address all correspondence to: hlwu@hnu.edu.cn

State Key Laboratory of Chemo/Biosensing and Chemometrics, College of Chemistry and Chemical Engineering, Hunan University, Changsha, China

References

[1] Yin XL, Wu HL, Gu HW, Hu Y, Wang L, Xia H, et al. Chemometrics-assisted high performance liquid chromatography-diode array detection strategy to solve varying interfering patterns from different chromatographic columns and sample matrices for beverage analysis. Journal of Chromatography A. 2016;**1435**:75-84

[2] Visky D, Haghedooren E, Dehouck P, Kovács Z, Kóczián K, Noszál B, et al. Facilitated column selection in pharmaceutical analyses using a simple column classification system. Journal of Chromatography A. 2006;**1101**(1):103-114

[3] Jandera P, Vyňuchalová K, Hájek T, Česla P, Vohralík G. Characterization of HPLC columns for two-dimensional LC× LC separations of phenolic acids and flavonoids. Journal of Chemometrics. 2008;**22**(3–4):203-217

[4] Haghedooren E, Farkas E, Kerner Á, Dragovic S, Noszál B, Hoogmartens J, et al. Effect of long-term storage and use on the properties of reversed-phase liquid chromatographic columns. Talanta. 2008;**76**(1):172-182

[5] Pérez RL, Escandar GM. Multivariate calibration-assisted high-performance liquid chromatography with dual UV and fluorimetric detection for the analysis of natural and synthetic sex hormones in environmental waters and sediments. Environmental Pollution. 2016;**209**:114-122

[6] Tan F, Tan C, Zhao A, Li M. Simultaneous determination of free amino acid content in tea infusions by using high-performance liquid chromatography with fluorescence detection coupled with alternating penalty trilinear decomposition algorithm. Journal of Agricultural and Food Chemistry. 2011;**59**(20):10839-10847

[7] Yin X-L, Wu H-L, Gu H-W, Zhang X-H, Sun Y-M, Hu Y, et al. Chemometrics-enhanced high performance liquid chromatography-diode array detection strategy for simultaneous determination of eight co-eluted compounds in ten kinds of Chinese teas using second-order calibration method based on alternating trilinear decomposition algorithm. Journal of Chromatography A. 2014;**1364**:151-162

[8] Wu HL, Li Y, Yu RQ. Recent developments of chemical multiway calibration methodologies with second-order or higher-order advantages. Journal of Chemometrics. 2014;**28**(5): 476-489

[9] Wang J-Y, Wu H-L, Chen Y, Sun Y-M, Yu Y-J, Zhang X-H, et al. Fast analysis of synthetic antioxidants in edible vegetable oil using trilinear component modeling of liquid chromatography–diode array detection data. Journal of Chromatography A. 2012;**1264**:63-71

[10] Braga JWB, Bottoli CB, Jardim IC, Goicoechea HC, Olivieri AC, Poppi RJ. Determination of pesticides and metabolites in wine by high performance liquid chromatography and second-order calibration methods. Journal of Chromatography A. 2007;**1148**(2): 200-210

[11] Fleming C, Kowalski B. Encyclopedia of Analytical Chemistry. New York: John Wiley& Sons Inc; 2000. pp. 9737-9764

[12] Malinowski ER. Factor Analysis in Chemistry, 3rd Ed. Technometrics. 2002;**36**(1):180-181

[13] Harshman RA. Foundations of the PARAFAC procedure: Models and conditions for an "explanatory" multi-model factor analysis. Ucla Working Papers in Phonetics. 1970;**16**

[14] Carroll JD, Pruzansky S, Kruskal JB. CANDELINC: A general approach to multidimensional analysis of many-way arrays with linear constraints on parameters. Psychometrika. 1980; **45**(1):3-24

[15] Wu H-L, Nie J-F, Yu Y-J, Yu R-Q. Multi-way chemometric methodologies and applications: A central summary of our research work. Analytica Chimica Acta. 2009;**650**(1):131-142

[16] Fleming CM, Kowalski BR. Second-order calibration and higher. Encyclopedia of Analytical Chemistry: Applications. Theory and Instrumentation. 2006

[17] Chen Z-P, Wu H-L, Jiang J-H, Li Y, Yu R-Q. A novel trilinear decomposition algorithm for second-order linear calibration. Chemometrics and Intelligent Laboratory Systems. 2000; **52**(1):75-86

[18] Xia AL, Wu HL, Fang DM, Ding YJ, Hu LQ, Yu RQ. Alternating penalty trilinear decomposition algorithm for second-order calibration with application to interference-free analysis of excitation–emission matrix fluorescence data. Journal of Chemometrics. 2005;**19**(2): 65-76

[19] Xia AL, Wu HL, Li SF, Zhu SH, Hu LQ, Yu RQ. Alternating penalty quadrilinear decomposition algorithm for an analysis of four-way data arrays. Journal of Chemometrics. 2007; **21**(3–4):133-144

[20] Law HG. Research methods for multimode data analysis. Praeger. 1984

[21] Wu HL, Shibukawa M, Oguma K. An alternating trilinear decomposition algorithm with application to calibration of HPLC–DAD for simultaneous determination of overlapped chlorinated aromatic hydrocarbons. Journal of Chemometrics. 1998;**12**(1):1-26

[22] Bro R, Kiers HA. A new efficient method for determining the number of components in PARAFAC models. Journal of Chemometrics. 2003;**17**(5):274-286

[23] Chen Z-P, Liu Z, Cao Y-Z, Yu R-Q. Efficient way to estimate the optimum number of factors for trilinear decomposition. Analytica Chimica Acta. 2001;**444**(2):295-307

[24] Li Y, Wu H-L, Qing X-D, Zuo Q, Chen Y, Yu R-Q. A novel method to estimate the chemical rank of three-way data for second-order calibration. Chemometrics and Intelligent Laboratory Systems. 2013;**127**:177-184

[25] Sanchez FC, Toft J, Van den Bogaert B, Massart D. Orthogonal projection approach applied to peak purity assessment. Analytical Chemistry. 1996;**68**(1):79-85

[26] Xie H-P, Jiang J-H, Long N, Shen G-L, Wu H-L, Yu R-Q. Estimation of chemical rank of a three-way array using a two-mode subspace comparison approach. Chemometrics and Intelligent Laboratory Systems. 2003;**66**(2):101-115

[27] Malinowski ER. Determination of the number of factors and the experimental error in a data matrix. Analytical Chemistry. 1977;**49**(4):612-617

[28] Xia A-L, Wu H-L, Zhang Y, Zhu S-H, Han Q-J, Yu R-Q. A novel efficient way to estimate the chemical rank of high-way data arrays. Analytica Chimica Acta. 2007;**598**(1):1-11

[29] Hu L-Q, Wu H-L, Jiang J-H, Han Q-J, Xia A-L, Yu R-Q. Estimating the chemical rank of three-way data arrays by a simple linear transform incorporating Monte Carlo simulation. Talanta. 2007;**71**(1):373-380

[30] Nie J-F, Wu H-L, Wang J-Y, Liu Y-J, Yu R-Q. The chemical rank estimation for excitation-emission matrix fluorescence data by region-based moving window subspace projection technique and Monte Carlo simulation. Chemometrics and Intelligent Laboratory Systems. 2010;**104**(2):271-280

[31] Amigo JM, Skov T, Bro R. ChroMATHography: Solving chromatographic issues with mathematical models and intuitive graphics. Chemical Reviews. 2010;**110**(8):4582-4605

[32] Zhang Y, Wu H-L, Xia A-L, Hu L-H, Zou H-F, Yu R-Q. Trilinear decomposition method applied to removal of three-dimensional background drift in comprehensive two-dimensional separation data. Journal of Chromatography A. 2007;**1167**(2):178-183

[33] Yu Y-J, Wu H-L, Fu H-Y, Zhao J, Li Y-N, Li S-F, et al. Chromatographic background drift correction coupled with parallel factor analysis to resolve coelution problems in three-dimensional chromatographic data: Quantification of eleven antibiotics in tap water samples by high-performance liquid chromatography coupled with a diode array detector. Journal of Chromatography A. 2013;**1302**:72-80

[34] Su Z-Y, Wu H-L, Liu Y-J, Xu H, Zhang J, Nie C-C, et al. Simultaneous determination of main effective constituents in traditional Chinese medicine Kudzuving root using HPLC-DAD coupled with second-order calibration based on alternating trilinear decomposition. Acat Chimica Sinica;**2011**(4):459-464

[35] Liu Y, Wu H, Zhu S, Kang C, Xu H, Su Z, et al. Rapid determination of costunolide and dehydrocostuslactone in human plasma sample and Chinese patent medicine Xiang Sha Yang Wei capsule using HPLC-DAD coupled with second-order calibration. Chinese Journal of Chemistry. 2012;**30**(5):1137-1143

[36] Ding Y, Wu H, Xia A, Cui H, Yu R. Simultaneous determination of two antituberculosis drugs using alternating fitting residue algorithm combined with HPLC-DAD. Journal of Analytical Science. 2008;**24**(1):1

[37] Zhao J, Wu H-L, Niu J-F, Yu Y-J, Yu L-L, Kang C, et al. Chemometric resolution of coeluting peaks of eleven antihypertensives from multiple classes in high performance

liquid chromatography: A comprehensive research in human serum, health product and Chinese patent medicine samples. Journal of Chromatography B. 2012;**902**:96-107

[38] Xiang SX, Kang C, Xie LX, Yin XL, Gu HW, Yu RQ. Fast quantitative analysis of four tyrosine kinase inhibitors in different human plasma samples using three-way calibration-assisted liquid chromatography with diode array detection. Journal of Separation Science. 2015;**38**(16):2781-2788

[39] Liu Z, Wu H-L, Li Y, Gu H-W, Yin X-L, Xie L-X, et al. Rapid and simultaneous determination of five vinca alkaloids in Catharanthus roseus and human serum using trilinear component modeling of liquid chromatography–diode array detection data. Journal of Chromatography B. 2016;**1026**:114-123

[40] Gu H-W, Wu H-L, Yin X-L, Li Y, Liu Y-J, Xia H, et al. Multi-targeted interference-free determination of ten β-blockers in human urine and plasma samples by alternating trilinear decomposition algorithm-assisted liquid chromatography–mass spectrometry in full scan mode: Comparison with multiple reaction monitoring. Analytica Chimica Acta. 2014;**848**:10-24

[41] Yin XL, Wu HL, Gu HW, Zhang XH, Sun YM, Hu Y, et al. Chemometrics-enhanced high performance liquid chromatography-diode array detection strategy for simultaneous determination of eight co-eluted compounds in ten kinds of Chinese teas using second-order calibration method based on alternating trilinear decomp. Journal of Chromatography A. 2014;**1364**:151

[42] Sun Y-M, Wu H-L, Wang J-Y, Liu Z, Zhai M, Yu R-Q. Simultaneous determination of eight flavonoids in propolis using chemometrics-assisted high performance liquid chromatography-diode array detection. Journal of Chromatography B. 2014;**962**:59-67

[43] Zhang X-H, Wu H-L, Wang J-Y, Tu D-Z, Kang C, Zhao J, et al. Fast HPLC-DAD quantification of nine polyphenols in honey by using second-order calibration method based on trilinear decomposition algorithm. Food Chemistry. 2013;**138**(1):62-69

[44] Yu Y-J, Wu H-L, Shao S-Z, Kang C, Zhao J, Wang Y, et al. Using second-order calibration method based on trilinear decomposition algorithms coupled with high performance liquid chromatography with diode array detector for determination of quinolones in honey samples. Talanta. 2011;**85**(3):1549-1559

[45] Liu Z, Wu HL, Xie LX, et al. Direct and interference-free determination of thirteen phenolic compounds in red wines using a chemometrics-assisted HPLC-DAD strategy for authentication of vintage year. Analytical Methods. 2017;**9**(22):3361-3374

[46] Wang T, Wu HL, Xie LX, Zhu L, Liu Z, Sun XD, et al. Fast and simultaneous determination of 12 polyphenols in apple peel and pulp by using chemometrics-assisted high-performance liquid chromatography with diode array detection. Journal of Separation Science. 2017;**40**(8):1651-1659

[47] Liu Z, Wu H-L, Xie L-X, Hu Y, Fang H, Sun X-D, et al. Chemometrics-enhanced liquid chromatography-full scan-mass spectrometry for interference-free analysis of multi-class

mycotoxins in complex cereal samples. Chemometrics and Intelligent Laboratory Systems. 2017;**160**:125-138

[48] Gu H-W, Wu H-L, Li S-S, Yin X-L, Hu Y, Xia H, et al. Chemometrics-enhanced full scan mode of liquid chromatography–mass spectrometry for the simultaneous determination of six co-eluted sulfonylurea-type oral antidiabetic agents in complex samples. Chemometrics and Intelligent Laboratory Systems. 2016;**155**:62-72

[49] Zhang Y, Wu H-L, Xia A-L, Han Q-J, Cui H, Yu R-Q. Interference-free determination of Sudan dyes in chilli foods using second-order calibration algorithms coupled with HPLC-DAD. Talanta. 2007;**72**(3):926-931

[50] Qing X-D, Wu H-L, Li Y-N, Nie C-C, Wang J-Y, Zhu S-H, et al. Simultaneous determination of pre-emergence herbicides in environmental samples using HPLC-DAD combined with second-order calibration based on self-weighted alternating trilinear decomposition algorithm. Analytical Methods. 2012;**4**(3):685-692

[51] Shen H-B, Yang J, Liu X-J, Chou K-C. Using supervised fuzzy clustering to predict protein structural classes. Biochemical and Biophysical Research Communications. 2005;**334**(2): 577-581

[52] Sun J, Wu H, Mo C, Lu J, Cui H, Yu R. Simultaneous determination of isomers of dihydroxybenzene using alternating trilinear decomposition algorithm combined with reversed-phase high performance liquid chromatography/diode array detection. Chinese Journal of Chromatography. 2002;**20**(5):385-389

[53] Lu J-Z, Wu H-L, Sun X-Y, Cui H, Sun J-Q, Yu R-Q. Simultaneous decomposition and determination of the complex dimethylphenol isomers by alternating trilinear decomposition algorithm combined with region selection. Chinese Journal of Analytical Chemistry. 2004;**32**(10):1278-1282